Electricity in fish research and management

Electricity in fish research and management

Theory and practice

W.R.C. Beaumont

SECOND EDITION

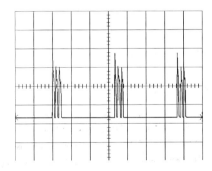
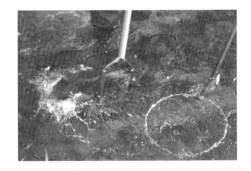

WILEY Blackwell

Library of Congress Cataloging-in-Publication Data

Names: Beaumont, W. R. C., author.
Title: Electricity in fish research and management : theory and practice / W.R.C. Beaumont.
Description: Second edition. | Hoboken : John Wiley & Sons Inc., 2016. | Includes bibliographical references.
Identifiers: LCCN 2015040512 | ISBN 9781118935583 (cloth)
Subjects: LCSH: Electric fishing. | Fishery management. | Fisheries–Research. |
   Electrophysiology–Research. | Fishes–Physiology–Research.
Classification: LCC SH344.6.E4 B43 2016 | DDC 639.2028–dc23 LC record available at
   http://lccn.loc.gov/2015040512

Contents

Acknowledgements

This book would not have been possible without the wealth of information that researchers have published over the years in their quest for understanding how electric fishing works (and the associated other uses of electricity in fisheries management). W. Gilbert Hartley (former UK Ministry of Agriculture, Fisheries and Food (MAFF) scientist) encouraged me to continue my studies and helped by donating his collected literature, much of which is now probably unobtainable. Alan Temple (US Fish & Wildlife Service National Conservation Training Center) provided good advice in the excellent US Fish and Wildlife electric fishing manual. Steve Miranda (Mississippi Cooperative Fish & Wildlife Research Unit, USA) gave an interesting insight into power transfer theory. Martin Rouen and the late Mike Lee gave simple advice regarding complex electronic queries. Graeme Peirson and the late Phil Hickley of the UK Environment Agency (EA) saw the need for further guidance material on electric fishing, and commissioned the EA *Guidelines for Electric Fishing Best Practice* which taught us all so much. The UK Game & Wildlife Conservation Trust published an earlier, smaller version of this manual. Will and Di Beaumont made helpful comments and together with Maddy gave support and encouragement during the long process of writing. Finally, I thank my mother for inspiring me.

CHAPTER 1

Introduction

Over the past 100 years, electricity, or he use of electrical theory, has become a vital tool for fishery research and management. It is used to divert fish from hazards, capture them for research or husbandry, count them as they migrate within rivers and anaesthetise/sedate them for easier handling or tagging. Due to the potential hazards associated with using 'free' electricity in an aquatic environment (to both operators and fish), there is a continuing need to develop and promote Best Practice Guidelines for its use. Also, recent advances in electronics have resulted in new equipment that allows far greater control and selection of the output from the equipment. This new availability of output settings makes the need for improved understanding and guidance on how the method works and suitable output settings even more necessary. Probably the commonest use for electricity in fish research and management is for capturing them; this is called **electric fishing** (or **electrofishing**). Even for this methodology, books about the method tend to be either too simplistic or collations of scientific papers. The aim of this book is to give comprehensive information about the technical and theoretical aspects of using electricity but in easily understood language and format. In addition, practical guidance gained from over 40 years of experience of using electricity for a variety of uses in a wide range of locations and conditions is also given. The book will concentrate on electric fishing but in doing so will explain the fundamental concepts that govern other uses of electricity in fish research and management.

There has recently been a greater awareness and concern for fish welfare while electric fishing, and this book emphasises the concept of promoting fish welfare above fish capture. Information on basic electric circuit theory, choice of equipment, output characteristics and use should enable adequate fish capture efficiency with minimum incidence and severity of fish damage. Information and guidance that enable users to have a good understanding of the factors that influence efficient equipment set-up and benign fish capture are fundamental to achieving these goals. The work presented in this document is intended to give the above guidance and is based on an earlier document on Best Practice

Electricity in Fish Research and Management: Theory and Practice, Second Edition. W.R.C. Beaumont.
© 2016 John Wiley & Sons, Ltd. Published 2016 by John Wiley & Sons, Ltd.

Guidelines for electric fishing (Beaumont *et al.* 2002) prepared for the UK Environment Agency. When Beaumont *et al.* (2002) was being written, it was hoped that it would be possible to lay down definitive rules and settings for use under a standard set of conditions and equipment specifications. Unfortunately, due to the wide variety of gears in use, water bodies in which they are used and range of operational requirements, it is not easy to categorically state what to use and where. Instead, it was decided that operators should have a good understanding of equipment and of the basic theory behind technique; this would allow them to set gear and output according to circumstances. This book is yet a further step along that road.

Overviews of electric fish screens, fish counters and the use of electricity to anaesthetise and sedate fish are also given.

The book is aimed at all who undertake sampling using electric fishing: professional practitioners such as government research scientists, under and post-graduate students and lay operators (water keepers etc.). Contrary to the comments of Smolian (1944) that 'at all costs electrofishing should not be allowed to develop into a method that allows any errand boy to be a fisherman', I hope that all who read this book will gain an understanding of at least the basic principles.

Recommendations from this work will include guidance on:
1 Output type and waveform
2 Frequency and power output
3 Anode size and shape, and cathode size and shape
4 Choice of options available regarding gear configuration (single anode, multi-anode, boom-mounted etc.)
5 Practical advice on using the equipment
6 Post-capture fish care.

Only core health and safety issues specifically associated with electric fishing will be addressed, as national, regional or local Codes of Practice or guidelines should deal with issues such as lifting and working near water, and so on.

Whilst this book is based on equipment and practice commonly used in Europe, the electrical principles described are universal and will apply to whatever type of equipment is being used.

CHAPTER 2

The history of electricity in fish research

In nature, certain fish have been using electricity for millennia, using it for defence (electric ray), sensory information (*marmoratus* sp.) and fish/prey capture (electric eel). Humans too have known about the physical effects and properties of an 'unknown force' (that we now call electricity) for a long time, but did not know the cause. Aristotle (350 BCE) mentions electric rays, and the Greek poet Claudius Claudianus (370–404 BCE) gives a very complete description of the effects of the electric ray on fishermen:

> "all that have touched it lie benumbed. … Should it carelessly swallow a piece of bait that hides a hook of bronze and feel the pull of the jagged barbs … emitting from its poisonous veins an effluence which spreads far and wide through the water. The poison's bane leaves the sea and creeps up the line; it will soon prove too much for the distant fisherman. The dread paralysing force rises above the water's level and climbing up the drooping line, passes down the jointed rod, and congeals, e'er he is even aware of it, the blood of the fisherman's victorious hand. He casts away his dangerous burden and lets go his rebel prey, returning home disarmed without his rod."

The *Encyclopaedia Americana* (Anon. 1918) notes that Arabs in the fifteenth century give the same name to lightning and electric rays, thus linking the two phenomena. Interestingly, Michael Faraday (1791–1867), one of the great electrical scientists (and by whose efforts electricity became practical for use in technology), also investigated the properties of the electric ray in the 1830s.

In addition, but not documented in written text, the South American electric eel has long been feared by native tribes. European explorers record the locals driving animals into streams to be immobilised by the eels, whereby they may be captured more easily. One of its names in South America translates as 'one who puts you to sleep', and there is documented evidence that it can use the 600 volts it can produce to knock out a human, as well as kill large caiman that are foolish enough to try to capture it (Nye 2014).

In terms of using electricity for fish research and management, the Italian Alessandro Volta made a significant step in 1791 when he published details of the first truly portable electricity supply, the voltaic pile or battery. There is some

Electricity in Fish Research and Management: Theory and Practice, Second Edition. W.R.C. Beaumont.
© 2016 John Wiley & Sons, Ltd. Published 2016 by John Wiley & Sons, Ltd.

debate over whether the Mesopotamians used a chemical battery over 2000 years ago, but Volta's was the first documented battery and was described some 150 years before the earlier artefacts were discovered. Until Volta's battery, electricity was produced by storing static electricity in 'Leyden jars'; these could be made to discharge and create short bursts of high voltage, direct current (DC) electricity. The galvanic pile was the first device able to produce a continuous and stable electrical current over a period of time.

It was an Englishman, Isham Baggs, who in 1863 patented the idea of using electricity from batteries for 'Paralysing fish, birds &c.' The patent was very thorough and covered what we would now know as standard electric fishing but also the use of electricity to immobilise fish after they have been hooked on a baited line in order to aid capture and minimise the chances of the fish escaping off the hook (a process that is now used in some tuna long-line fisheries) and the use of polarising glasses, made from tourmaline, in order to better observe the position of the fish in the water (Baggs 1863).

Having discovered the effect that electricity had on fish, scientists then began a long process of trying to understand the reason for the effects on fish (and other animals) and the cause of the effects. Mach (1875) described the orientation and movement of fish in an electric field, discovering that fish turn towards an anode (galvanotropism). Herman (1885) also reports orientation in an electrical current and movement towards anodes (galvanotropism and taxis). This was confirmed by Blasius and Schweitzer (1893) and Nagel (1895). Herman and Matthias (1886) reported that fish experienced 'discomfort' in a reversed field (cathodic repulsion). Subsequent to these early studies, a considerable amount of research was carried out on the physiological and practical factors that determined the reaction of the fish. Much of this early research was carried out in Germany, France and America, but Russia and Japan also published information relating to electric fishing. Loeb and Maxwell (1896) considered the response was involuntary, with current flowing through the central nervous system and affecting the motorneurones and flexor–extensor muscle systems. Much of this early work, however, had poorly described experimental and output settings, was not particularly applicable to understanding the process in the natural environment and was also often published in obscure journals. More recently, greater attention to detail has improved the description of electrical parameters; however, it is still common for papers not to document waveforms properly (e.g. Van Zee *et al.* 1996) or note whether water conductivity is specific or ambient.

All of the early research on using electricity for fisheries research used DC fields, but in 1902 a Frenchman, Professor Stéphane Leduc, described using a pulsing waveform to reduce the power demand of electric fishing. This pulsed waveform was a square wave with a frequency of between 20 and 200 Hz and a 0.05–0.005 s duty cycle (described by May 1911). It was derived from interrupting the voltage from the galvanic cell batteries in use at the time, and for some while the waveform was called 'Leduc's current'.

Much of the early research on the practical uses of electricity for fisheries management was directed at using electric fields to guide fish or as fish barriers. In 1917, H.T. Burkey was awarded the first of a series of patents for an electric fish screen, and this is probably the first true use of electricity for fisheries management.

The widespread use of electricity to catch fish and use the method for fisheries research and management probably started around the early 1930s (Holzer 1932) with a range of different equipment designs being described by various authors over the succeeding years.

Since the 1940s, many studies have described the practical uses of the method for fisheries research and management and also increased our understanding of how the process works. Each country seems to have had a core of researchers who specialised in furthering our knowledge.

Researchers in America published some of the first details of practical designs for electric fishing equipment and also studies on how the process affected fish. Researchers also demonstrated that the effect on the fish was independent of the central nervous system (as freshly killed fish that had had their spines or spinal cords removed still reacted to an electric field and 'swam' towards the anode) and also documented the differences in effect between DC and pulsed DC (pDC) waveforms (Haskell *et al.* 1954). More recently, researchers have progressed our understanding of factors affecting the effect on fish and the damage that can also be caused if used at inappropriate settings.

In Europe, several countries researched the practical use and physiology of the reaction. France in particular made big advances in the 1960s in researching both the physiology and practical use of the method. Latterly, the Food and Agriculture Organization (FAO) and the European Inland Fisheries Advisory Commission (EIFAC) have taken a lead role in collating research from Europe and identifying areas where more research is required.

Russia carried out a considerable amount of research, much of it focussed on the commercial (both freshwater and marine) aspects of the method, but also on the theoretical aspects (e.g. Sternin *et al.* 1976). The work was often unavailable to Western researchers, however, and whilst some key works have been translated into English they are still a rare find. The Russian work can be very mathematical and often relies on mathematical extrapolation of quite advanced theoretic concepts rather than empirical evaluation.

Japan also was involved with research on electric fields from the 1950s, but again much of the work was not accessible to Western researchers.

In the 1950s, Lethlean described using electrical principles to detect and count fish (Lethlean 1954). The method uses the change in the electrical resistance of the water caused by the presence of a fish in the water to register a count. It is used principally for counting large (>500 mm) migratory (anadromous) fish but can be used to count smaller (potamodromous) fish.

Whilst Isham Baggs' patent mentions 'paralysing' fish, it was not until the early 1930s that the idea of actively using electricity to anaesthetise fish started to be explored. Many scientific procedures and management operations are made easier if the fish are sedated. Even such simple tasks as measuring and weighing fish are difficult if the fish are wriggling. Conventional chemical sedation leaves residues in fish that may have implications if the fish is subsequently eaten by humans or other animals; electricity does not have this problem. However, comprehensive research is still lacking in the settings required for this use of electricity.

CHAPTER 3

Electric fishing

Electric fishing (or **electrofishing**) is the term given to a number of very different sampling methods. All have in common the utilisation of the reaction of fish to electrical fields in water for facilitating capture (Hartley 1980a, Pusey *et al.* 1998). At its most basic, electric fishing can be described as 'the application of an electric field into water in order to incapacitate fish, thus rendering them easier to catch'.

Despite over 100 years of study, the exact nature by which these effects are caused is still a matter of some debate (Sharber & Black 1999, cf. Kolz 1989, Reynolds *et al.* 1988, Snyder 2003). The basic principle is that the electrical field stimulates a muscular reaction (either involving the central and/or autonomic nervous system or not) resulting in the characteristic behaviour and immobilisation of the fish.

Two views on the underlying cause of the effect predominate, the 'Biarritz Paradigm' and the 'Bozeman Paradigm'. The former, which was proposed by Lamarque (1967, 1990) but also includes the principles underlying Kolz's Power Transfer Theory (Kolz 1989), considers the phenomenon to be a reaction to electrostimulation of both the central nervous system (CNS) and autonomic nervous system and the direct response of the muscles of the fish (i.e. a reflex response) (Sharber & Black 1999). In 1999, Sharber and Black (1999) proposed an alternative theory, the Bozeman Paradigm. In this theory the fish response is basically that of electrically induced epilepsy, and when the electrical stimulation overwhelms the CNS the (epileptic) seizures occur.

Little external research has been carried out on Sharber and Black's epilepsy theory, but many studies have either supported or refuted the theory regarding the role of the fish's nervous system in determining the effect. Haskell *et al.* (1954) considered that the effect was independent of the CNS, as freshly killed fish that had had their spines removed or been pithed still reacted to an electric field and 'swam' towards the anode. Flux (1967) also found that dead fish responded to an alternating current (AC) voltage gradient and attributed this to Vibert's (1963) assertion that, for a direct current (DC) waveform, in tetanus

---

Electricity in Fish Research and Management: Theory and Practice, Second Edition. W.R.C. Beaumont.
© 2016 John Wiley & Sons, Ltd. Published 2016 by John Wiley & Sons, Ltd.

(where the fish's muscles go into spasm and are in a cramped state) the electricity is acting directly on the fish muscles (i.e. no CNS reaction). Sternin *et al.* (1976), quoting work by Danyulite and Malyukina (1967), also considered that their work disproved the role of neural action in stimulating the fish muscles and proved that electrotaxis is possible without participation of the brain. However, Stewart (1990), working on marine fish species, observed that a pulsed DC (pDC) waveform acted directly on the fish muscles, with the fish muscle reacting to each pulse, and considered that the electrical waveform was working in parallel with the nervous system to activate the fish's muscle system.

Given the wide variety of research findings on the fundamental cause of the effect, for the time being we need to accept that the underlying principles behind the response are not proven.

It is generally accepted by all researchers that it is the current density (amps/cm^2), which can also be expressed as the power density (watts/cm^3), which is the principle determinant of the behavioural response. The magnitude of the current density that the fish experience is governed by the applied voltage (and thus the voltage gradient (E) in the water), the conductivity of the water and the electrical conductivity of the fish.

In addition, it is possible that the fish skin acts in a way whereby electricity is more easily transmitted into the fish when there is a change in voltage potential around the fish: it is thought that this is due to the fish skin acting as a capacitor. This can be seen in experiments where fish have been put in water that has a gradually increasing DC voltage applied; eventually, when the voltage gradient is high enough, the fish will react, but the same reaction will occur at much lower voltage gradients if the voltage is switched off, then on. This would also explain the different fish reactions between DC and pDC waveforms (see Section 4.2) and also the cause of the increasing injury rates as pulse frequency increases (see Section 4.2.3.1).

The factors that affect the effectiveness of electrofishing include:
- Electrical waveform type
  - Including pulse shape, pulse frequency and pulse width
- Electrode design
- Water conductivity
- Fish conductivity
- Streambed conductivity and substratum type/topography
- Water temperature
- Fish size
- Time of day
- Fish species
- Water clarity
- Water width and depth
- Operator skill.

Within the user community, the lack of adequate information regarding these factors has resulted in electric fishing being regarded as an art rather than a science

(Kolz 1989). This lack of fundamental perception is encapsulated by the once-common practice of referring to the pulse box as the 'magic box'. Whilst it is possible to capture fish without knowing how the technique works, knowledge of the fundamentals will enhance catch efficiency and help reduce some of the drawbacks concerning injury, as mentioned here. Knowledge of the basic electrical principles will also allow equipment to be calibrated to produce similar fish capture probabilities and thus improve standardisation between sampling in different locations.

Electric fishing has advantages over many of the other fish survey methods available (e.g. snorkelling, netting and bankside observation) regarding the composition of the species captured. Wiley and Tsai (1983) found that electric fishing produced better and more consistent results than seines, gave better population estimates, caught larger fish than seine netting and caught more fish by total weight. Capture rates can also be much higher; Growns *et al.* (1996) found capture rates nearly 30 times greater for electric fishing compared to gill netting with twice as many species captured. Likewise, Pugh and Schramm (1998) found that electric fishing was far more cost-effective than hoop nets, with hoop netting only catching two species compared with 19 by electric fishing. Snorkelling has also been suggested as an alternative to electric fishing; again, however, sampling efficiency is lower and results are more variable than for electric fishing (Cunjak *et al.* 1988, Hayes & Baird 1994). Shallow areas with high velocities and coarse substrate are particularly difficult for fish assessment by snorkelling (Heggenes *et al.* 1990). Observing fish from the bankside has also been tried as a method of estimating fish species. Whilst good agreement between observations and depletion electric fishing estimates has been obtained for trout fry, correlations between bankside visual counts and adult numbers were low (Bozek & Rahel 1991). An additional advantage of electric fishing is that it does not require prior preparation of the site (with consequent delay and disturbance of the fish to be investigated), and the requirements in terms of manpower are small when compared with many of the other methods.

Electric fishing is not, however, a universal success. Researchers have found drawbacks with the method regarding assessing species assemblage patterns (Pusey *et al.* 1998), post-fishing induced movement (Nordwall 1999), immune system suppression (VanderKooi *et al.* 2001) and elevated blood plasma–cortisol levels (Beaumont *et al.* 2000). The method also has potential to cause injury (both physical and physiological) and, in extreme circumstances, death to the fish. Physical damage can occur when fish muscles react so strongly to the electric field that they break the fish's spine or ribs. The problem is not simply one of too high a voltage gradient, as Ruppert and Muth (1997) found that injuries occurred at field intensities lower than the threshold required even for narcosis. The most common injury observed is 'burn' or 'brand' marks (Figure 3.1). These brands can also take the form of making one quarter of the fish dark (Figure 3.2). Sometimes two quarters are coloured, and these tend to be opposing quarters (e.g. front left and rear right). These can be caused by melanophore discharge resulting from too

Figure 3.1 Example of an electrode 'burn' (indicated by arrow) on an Atlantic salmon.

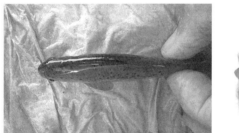

Figure 3.2 Example of an electrode 'Harlequin burn' on an Atlantic salmon parr.

Figure 3.3 Example of a spinal haematoma (indicated by arrow) caused by electric fishing on a rainbow trout.

close contact with (but not necessarily touching) the electrode, and they can be indicative of underlying spinal nerve damage. Spinal haematomas (Figure 3.3) and broken spines and ribs are caused by the electrical stimulation causing over-vigorous flexing of the muscles around the spine. Snyder (1995, 2003) gives a comprehensive review of pre-2000 research findings relating to fish damage.

The problems of fish injury and mortality have been the subjects of much debate and research since the 1940s. However, the literature is complex, often inconsistent and sometimes contradictory (Snyder 2003, Solomon 1999). Evidence exists that different species react differently to the technique (Pusey *et al.* 1998), injuries to captured fish can range from 0 to 90% (Snyder 2003) and even within the same species injury rates can vary. Whilst a definitive reason for many of these differences between results has, as yet, still to be unequivocally proven, two reasons predominate. One is that many studies have been carried out in different environments (i.e. in either laboratory or river conditions). In the laboratory set-up, conditions are dramatically simplified compared with natural conditions, and the electric field is homogeneous. Conversely, in the river the conditions are constantly changing and the fish are in different orientations and moving at various speeds, thus the electrical conditions are extremely variable (heterogeneous). Therefore, attempting to apply the results from one system to the other tends to throw up contradictions. Experimental results with fixed electric systems in running water can produce comparable results that can be analysed, but straightforward fishing is rarely an exact science. Another reason for many of the discrepancies between differing findings is that there is considerable doubt regarding the waveforms being used, with researchers often thinking they are using one waveform, but in fact are using another (Hill & Willis 1994, Van Zee *et al.* 1996). Thus, results for allegedly one waveform may be in fact for another.

However, whilst injuries undoubtedly do occur, they should be put into context regarding the population and natural mortality dynamics of the fish. Schill and Beland (1995) considered that, at a population level, even high electric fishing mortality rates have limited impact on species with high natural mortality rates. Pusey *et al.* (1998) found that fishing mortality, for a range of species, was generally less than 5% – this compares with annual natural mortality rates of >80% for many juvenile salmonid species. In addition, although every effort should be made not to harm the fish, often fish are able to recover from injuries with little long-term effect. Schill and Elle (2000) found that even when fish were subjected to DC and pDC electric fields intense enough to produce haemorrhage in ~80% of study fish, the injuries healed and did not represent a long-term mortality or health risk to the fish.

Non-electrical damage also needs addressing, including treading on incapacitated fish, damage from contact with the net frame, abrasion of fish from using knotted mesh nets, anoxia from being kept out of water too long and overcrowding in holding bins. Damage to non-fish species should also be considered of all animals (invertebrates, reptiles, birds and mammals) living in the water and being affected by the electric field.

These disadvantages can be reduced to negligible levels by the choice of appropriate method, suitable training for personnel and the experienced use of the apparatus (Hartley 1975). For healthy populations and with optimal (for fish welfare) gear set-up, the method can be considered acceptable given that 'normal' fishing mortality rates should have limited impact at a population level (Schill & Beland 1995). It should be noted that all removal sampling methodology is likely to result in some mortality. Even angling can produce mortality effects in fish, with Brobbel *et al.* (1996) reporting 12% mortality and Gargan *et al.* (2015) reporting 45% mortality (of lure-caught Atlantic salmon) after angling, probably due to lactic acid build-up and release in the muscles (Wood *et al.* 1983) which is enhanced after air exposure (Ferguson & Tufts 1992). Bouck and Ball (1966) also found that seining, angling and electroshock all produced adverse effects on rainbow trout blood chemistry and increased mortality, with the highest mortality rates being found for capture by angling.

The acknowledged problems associated with electric fishing–induced fish injury and mortality should be considered before deciding to use the technique, and users of the method have an obligation to select an appropriate and humane system of electric fishing. This method selection should encompass the choice of equipment for both different fish species and differing environments. In addition, users have a duty of care with regard to minimising stress and injury to fish during essential studies on fish populations. Ideally, the choice of electric fishing system should aim to achieve the optimum combination of both capture efficiency and fish welfare. It is recommended, however, that fish welfare should take priority over fish capture.

As an example of what can be achieved in practice, the author has been involved for several years with a project where every year approximately 10,000 juvenile Atlantic salmon parr are captured using electric fishing (with subsequent passive integrated transponder (PIT) tag implantation). By tailoring output settings and fishing method, capture rates of 40–70% are achieved and immediate post-capture mortality is less than 2%.

Lack of standardization and poor descriptions of equipment set-up make much of the research articles published during the early use of electric fishing of limited value for getting a historic perspective about fish stocks (Bohlin *et al.* 1989). This is particularly so with regard to population density data. Even today, it is common to see sample methods simply described as 'electric fishing'. Method descriptions should include waveform type, voltage and amperage values, anode diameter, cathode size, water conductivity (and whether specific or ambient), water temperature, river conditions (slow/fast, substrate type) and so on. Whilst it would be impractical to insist that all users of electric fishing use a standard equipment set-up, a consistent approach to electric fishing gear selection and output, together with a good description of those, would allow better comparison and standardisation of sampling between the same sites fished at different times or between sites in differing locations.

3.1 Health and safety

The thought of creating an electric field in water and then having operators wade in that water is enough to give most health and safety (H&S) personnel apoplexy! This is particularly so when H&S officers do not have a full appreciation of the method. It must be accepted, though, that electric fishing has the potential to be hazardous to operators. However, electric fishing has many H&S advantages over other methods available to fishery workers for capturing fish, and it has been widely used throughout the world for research and management for the past 60–70 years. It has many advantages over other sampling methods and the number of operators required is low, with just two people as a minimum for safety reasons.

There is some risk, however, as the electrical current used is intrinsically lethal. Safety guidelines (Anon. 2009) indicate that as little as 50–150 mA are enough to cause an 'extremely painful shock, respiratory arrest (breathing stops) and severe muscle contractions. Flexor muscles may cause holding on; extensor muscles may cause intense pushing away. Death is possible'. At amperages of 1–4 A, a value that is commonly used in moderate-conductivity water bodies (>300 μScm^{-1}), 'Ventricular fibrillation (heart pumping action not rhythmic) occurs. Muscles contract; nerve damage occurs. Death is likely'.

The potential for accidents increases when untrained operators are using the equipment and particularly so when large numbers of people are involved (Hartley 1975). In India, several deaths have occurred from subsistence fishermen hearing of the method and throwing wires on domestic overhead electrical supply lines (Balachandran *et al.* 2013). In Europe, to the author's knowledge, serious accidents have not occurred, and the fact that nobody has been injured, given some of the incredibly poor apparatus handled by untrained amateurs, in the past 40 years is notable. This is even more surprising given the once 'common' method of checking the output of the equipment by putting both hands into the water to determine whether the current was flowing (author's experience, Smolian 1944). There is therefore no room for complacency, and discussions at the author's training classes indicate that minor shocks are not uncommon. A survey in the United States found that up to 91% of fisheries personnel had been shocked whilst using the equipment (Lazauski & Malvestuto 1990). Allen-Gil (2000) gives a good example of how easy it is for accidents to happen: a simple problem of the insulating plastic missing from the metal wire handle of a bucket being used to hold captured fish resulting in the bucket carrier being immobilised when an anode touched the metal bucket handle.

Most countries will have specific H&S regulations that cover the use of the method. It is also important to ensure that these national criteria are being adhered to (Hickley & Millwood 1990). In some countries, additional regional or institutional H&S regulation may also cover construction criteria for the equipment (e.g. the UK Environment Agency). Care needs to be taken, however, that these construction criteria do not create a straitjacket that does not allow equipment to be varied to cope with the wide range of operating conditions that are found when electric fishing. It should be remembered that the principle danger from (correctly performed) electric fishing is from non-electrical hazards (Hartley 1975).

In Europe, the European Committee for Standardization (CEN) regulations cover sampling protocols, equipment specification and safety aspects (Anon. 2003a). These regulations were drafted by a committee of working fish ecologists with a good appreciation of the use of the equipment and, as a result, are complete but not too prescriptive. Other countries – and, in the case of the United States, many individual states – have their own criteria and regulations. These can cover 'simply' human H&S issues but also fish welfare issues resulting from different practices, equipment or electrical waveforms used. The following is a brief overview of the principle dangers of electric fishing to operators. All electric fishing operations and organisations using electric fishing should have individual and more extensive H&S guidelines.

No one should be in close proximity with energised electrodes if they have a history of cardiac problems. Severe electric shocks can cause distortion of the heart's rhythm and/or respiratory arrest. Likewise, stress-induced asthma can be triggered by an electric shock. It is recommended that **at least** one person in each electric team be trained in cardiopulmonary resuscitation (CPR) techniques.

All equipment used must be in good condition and should be suitable for the purpose of electric fishing. It should be regularly checked by a competent person and visually checked after each use. Faults must be reported, and faulty equipment must not be used.

There are four main hazards associated with electric fishing: electric shock, drowning, tripping/falling and trauma injuries.

3.1.1 Electric shock

The severity of electric shock is related to the amplitude of the current, the duration of the shock and the waveform. Direct current (DC) causes a severe shock only when the current is made or broken, but not when the current is steady. By contrast, alternating current (AC) will produce a continuous painful shock. It also requires ~3 times more DC than AC for a lethal shock. It should be noted that very low-amplitude shocks can give rise to severe effects, with breathing and heart performance impacted at 20–50 mA and fatal heart irregularities occurring at >50 mA: these levels should be set against the common use of electric fishing currents in excess of 2 A and in some cases up to 30 A.

In electric fishing, these dangers are exacerbated by the fact that the skin may be wet when a shock is received, reducing the electrical resistance of a 'dry' body by 10-fold, and thus increasing the current (amps) experienced by the body (amps = volts ÷ resistance). In addition, the 50–60 Hz pDC currents commonly

Table 3.1 The impact on humans of different current amplitudes.

Current level	Human reaction
1 mA	Just a faint tingle.
5 mA	Slight shock felt. Disturbing but not painful. Most people can 'let go'. However, strong involuntary movements can cause injuries.
6–25 mA (women), 9–30 mA (men)	Painful shock. Muscular control is lost. This is the range where 'freezing currents' start. It may not be possible to 'let go'.
50–150 mA	Extremely painful shock, respiratory arrest (breathing stops), severe muscle contractions. Flexor muscles may cause holding on; extensor muscles may cause intense pushing away. Death is possible.
1000–4300 mA (1–4.3 A)	Ventricular fibrillation (heart pumping action not rhythmic) occurs. Muscles contract; nerve damage occurs. Death is likely.
10,000 mA (10 A)	Cardiac arrest and severe burns occur. Death is probable.

Source: Anonymous (2009).

used when electric fishing are exactly the frequency needed to send the heart muscles into seizure (Anon. 2009). Table 3.1 gives the likely impacts on the human body for a range of current levels.

The most dangerous situation in electric fishing is when an energized electrode is out of the water. In this situation, there is a very real risk of severe electric shock if the electrode touches someone. Due to the possibility of anode switch failure, electrodes should be considered as 'live' whenever connected to a running generator or battery power supply. Shocks from contact with the water near live electrodes (provided contact is not directly on the electrode) are less severe due to the dissipation of the electric current in the water. Shocks that may occur from touching faulty equipment should not occur provided adequate servicing of gear is carried out and operators are vigilant for damage to equipment and cables.

In several countries, the use of electrically insulating 'linesmen's gloves' is mandatory in order to minimize the risk of operators' hands touching the water. See Section 5.6.2 for more detail of the use of these.

Electric shock can give rise to three major symptoms:

Ventricular fibrillation. This is the uncoordinated contraction of the ventricular muscle fibres of the heart. The greatest risk of ventricular fibrillation occurs when an electric shock is received and the path of the current is through the chest (e.g. between two arms). The heart's natural rhythm is stopped and replaced by an irregular quivering. This is extremely dangerous and, unless correctional steps are taken immediately, death can occur in minutes. CPR can be used to stabilize the patient until they can be defibrillated, but this is only a temporary tactic and is unlikely to restore the pulse by itself.

Respiratory arrest. Either electric shock or blows to the head can cause this. The control centre for respiration is contained at the base of the skull and can be deactivated by an electric shock or blow to that area. CPR or artificial respiration can help in certain cases.

Asphyxia. If operators are continuously shocked during electric fishing, the chest muscles can contract and not release, causing asphyxia.

3.1.2 Drowning

Although cardiac and/or nervous system injury resulting from the electric current is considered the most likely danger from electric shocks during fishing, drowning is also a significant danger. Electric shock can impair the swimming ability of operators, and operators should always wear life jackets or buoyancy aids where there is a risk of being shocked or falling whilst in deep or fast flowing water.

3.1.3 Tripping or falling

Movement on riverbeds and boats can be made difficult by slippery surfaces. Always try to ensure that nonslip footwear is worn whenever possible. Metal-studded boots are very effective with slippery algae-covered rock, but care is needed that the studs do not go all the way through the boots and come into direct contact with the wearer. Operators should move at a pace that is consistent with conditions underfoot. Be aware of trip hazards such as cables and ropes on the ground.

3.1.4 Trauma

Net handles can inflict significant injury to nearby members of fishing teams. In particular, a net handle poked into an eye can cause grievous injury. Care should be taken when netting, and other members of the team (cable gaffers, bucket carriers etc.) should stay clear of the net operators.

3.2 General issues

Electric fishing can be a very physical activity, and operators need to ensure they have adequate fitness for the task. Moving electrodes and nets through water can rapidly tire arms that are not used to physical exertion. Wading in even moderately deep flowing water can be tiring, electric fishing equipment can be heavy as can the buckets and bins of water that hold the fish. Concentration levels needed when fishing (particularly when capturing small fast moving fish) are high and can also be tiring. Operators should make sure they keep adequately hydrated and fed, and have regular rests.

CHAPTER 4

Electrical terms

The principles and terms described in this chapter are universal for electric fishing, fish screens, resistivity counters and electroanaesthesia. However, the descriptions will concentrate on their application with reference to electric fishing.

The physical parameters of electricity can be split into six principle components. Being invisible, electrical parameters can be difficult to conceptualise. For this reason, the descriptions here will also use, where appropriate, an analogy of water within a pipe to try to help explain more clearly the parameters.

1 **Circuit**: The closed pathway that the electrical current travels through. In 'normal' electrical circuits, this would consist of the wires and components within the apparatus; in electric fishing, the 'closed' pathway also includes the water that the electric field is discharged into.

2 **Voltage**: The amount of potential energy between two points on a circuit. It is measured in volts (V), and one volt is the potential energy difference between two points that will impart one joule of energy per coulomb of charge that passes through it. It can be likened to the pressure of water within a pipe.

  (a) **Voltage gradient**: The difference in voltage between two points (in water). Measured in volts per centimetre (V.cm^{-1}).

  (b) **Voltage waveform**: The shape of the voltage when plotted on a graph over time. For electric fishing, three principle waveforms are used:

   b.1. *Alternating current (AC)*: Voltage alternates from positive to negative. In electric fishing, this waveform should **only** be used as an output from generators to control boxes.

   b.2. *Direct current (DC)*: Steady positive voltage (as from a battery).

   b.3. *Pulsed direct current (pDC)*: Pulses of DC voltage.

3 **Current**: The rate of flow of charge in a circuit (i.e. coulombs per second). The symbol for electric current is 'I' (standing for 'intensity' of current), but it

Electricity in Fish Research and Management: Theory and Practice, Second Edition. W.R.C. Beaumont.
© 2016 John Wiley & Sons, Ltd. Published 2016 by John Wiley & Sons, Ltd.

is measured in amperes (with the symbol A). It can be likened to the volume of water passing through a pipe.

   (a) **Current density**: The ratio of the current to the cross-sectional area of its path. Measured in Amps.cm^{-2}, its symbol is J.

4 **Power**: The rate at which electrical energy is transferred by an electric circuit. It can be likened to the work the water in a pipe could do (turning a turbine etc.). It can be calculated in three ways:

$$Power = volts \times amps$$

$$Power = amps^2 \times resistance$$

$$Power = volts^2 \div resistance.$$

   • For DC and pDC waveforms, power is measured in watts. For AC waveforms or DC/pDC waveforms produced from AC generators, a correction factor (power factor (PF)) may be required, and the units of measurement are volt-amps (VA).

5 **Resistance**: A measure of the difficulty that the electrical force encounters in passing a current through the medium in which it is contained. Measured in ohms (Ω). It can be likened to the amount of difficulty or friction water has in passing through different diameter pipes.

   (a) **Equivalent resistance (EqR)**: The resistance of an electrode and the water in which it is present; a function of the resistance (see above) of the electrode material, together with its shape and the resistance of the water.

6 **Conductance**: The measure of the ease with which the electricity flows through the substance in which it is contained. It is the reciprocal of electrical resistance ($1/\Omega$). The current SI unit of measurement is Siemens (S), symbolised by ó, but it was previously measured in mho.

   (a) **Conductivity**: The ability of a unit volume of matter to conduct electricity. Measured in Siemens per meter (S.m^{-1}) but in water usually micro-Siemens per centimetre (μS.cm^{-1}) with 100 μS.cm^{-1} = 0.0100 S.m^{-1}.

Items 2, 3 and 5 can be calculated provided the other two are known, using the relationship diagram in Figure 4.1.

Figure 4.1 Relationship diagram for calculating volts, amps and resistance. Removing and placing the factor to be calculated to the left of the triangle leave the calculation maths from the other two factors.

4.1 Circuit

In 'normal' electrical circuits, this would consist of the wires and components within the apparatus; in electric fishing, the closed pathway also includes the water into which the electric field is discharged. The fact that the circuit includes the water body also introduces aspects such as boundary layers that confine the water body. These boundaries are principally the surface of the water (which forms a hard, defined boundary) and the area at the lower limit of the electrical effect. In normal electric fishing in streams and shallow water, this will be the bed of the river or water body; due to the water infiltration into this bed, the boundary is not as defined as with the water surface, and the conductivity of the bed material will also affect the circuit characteristics. Weed and other electrically conductive or insulating debris within the water body will also affect the circuit.

4.2 Voltage

The field lines of the voltage and current around the electrodes conform to the pattern shown in Figure 4.2. The voltage does not 'flow' from one electrode to another, and under isolated conditions it can be described as a series of spheres of progressively smaller electrical potential around the electrode (similar to the rings or layers of an onion, with each layer having successively lower voltages as the distance from the centre increases). When the fish are aligned along the

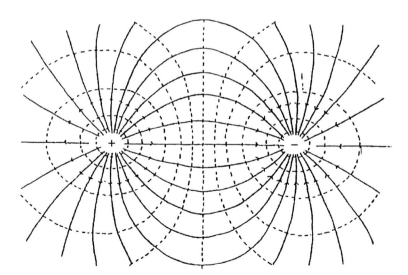

Figure 4.2 Generalised pattern of voltage gradient (dashed lines) and current (solid lines) around two similar sized but opposite polarity electrodes in close proximity in a conductive medium.

current lines, they will experience the greatest voltage potential. When they are aligned along the voltage lines, they should experience the least head-to-tail voltage difference (but will experience some lateral voltage gradient across their body).

The literature describes four classic zones of effect of an electric field, each occurring at differing distances away from the source (Vibert 1967, Regis *et al.* 1981, Snyder 1995). Some zones are common to all electric current types, and some are specific to one type. Attraction and narcosis zones are for anodic (or positive) direct current (DC) or pulsed direct current (pDC) waveforms only (see Section 4.3). Cathodic (or negative) DC or pDC induces a repelling field, causing the fish to turn away from the cathode and towards the anode, a reaction often used for fish exclusion screens. However, tetanising effects can still occur in high-cathode fields.

Common behaviour zones of an electrical field are as follows.

1 The **indifference zone** is the area where the electric field has no influence upon the fish.

2 The **repulsion** or **fright zone** occurs on the periphery of the field; the fish feels the field, but it is not intense enough to physiologically attract the fish. The fish instead reacts as to any reactive stimulus; this may include escaping or seeking refuge (hiding in weed beds or burrowing into the bottom, depending on the species). Intelligent use in operating the anode can limit a fish's probability of encountering this zone.

3 The **attraction zone** (DC and pDC only) is the critical area where the fish is drawn towards the electrode. This occurs due to either anodic taxis (swimming towards the anode driven by the electric field's effect on the fish's central nervous system (CNS) and/or muscles) or forced swimming (involuntary swimming caused by direct effect by the electric field on the autonomic nervous system). In the latter case, swimming motions often correspond with the initial switching of DC and the pulse rate of pDC. Fishing equipment and output should seek to maximise this zone.

4 The **narcosis zone** is where, in normal DC and pDC fields, immobilisation of the fish occurs. In this state of narcosis, the fish muscles are relaxed and the fish still breathes (albeit at a reduced level). When removed from this narcotising zone, fish should recover instantly and behave in a relatively normal manner.

5 The **tetanus zone** is the region where immobilisation (from AC, pDC and some DC fields) of the fish occurs. In AC, pDC and very high DC fields, fish in this zone have their muscles under tension and respiratory function ceases. Fish may require several minutes to recover from this state. Tetanus can harm fish, and thus this zone should be minimised in gear design and selection and fish removed quickly from it.

Sternin *et al.* (1976) describes a very comprehensive seven-point system of reaction types with additional supplementary notation allowing anode, cathode and

AC effects and the percentage of fish that show the reaction (in tests) to be represented. Generally, however, they consider that a three-stage classification (of I Excitement, II Electrotaxis and III Shock) is sufficient when a detailed classification is not needed.

Voltage can be described and measured in a number of ways.

Peak voltage: Zero to maximum (V_{pk})

Peak-to-peak voltage: Minimum to maximum (V_{pp})

Root mean square (RMS) voltage: This quantifies the equivalent steady DC voltage that would transfer the same power into the water (V_{rms}). In simple terms, it can be likened to average voltage over time.

Peak voltage will measure the maximum voltage attained, and peak-to-peak is used for voltages that have a negative component to them; this includes AC but also some poorly formed (often purportedly pDC) voltage waveforms. For steady DC, both V_{pk} and V_{rms} methods will give the same reading; and V_{pp} equals zero (Figure 4.3). For pulsed voltages, however, V_{pk} and V_{rms} will give a different answer (Figure 4.4). Peak voltage will measure the maximum voltage attained by the pulse, while the RMS value will be lower. Most standard voltmeters can measure either steady DC voltage or AC voltage; if used on pDC waveforms, they will give V_{rms} values. Only specialised digital volt meters (DVMs) and oscilloscopes can measure the peak (and peak-to-peak) voltage of pulsed currents.

Figure 4.3 Measuring DC voltage.

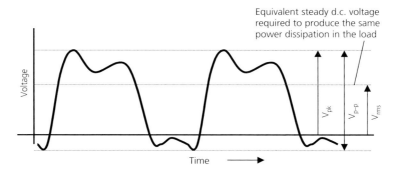

Figure 4.4 Measuring pDC voltages (including where negative component present).

4.2.1 Voltage gradient

When an electric current is passed through water from one point (electrode) to another, it dissipates and can, with sensitive enough equipment, be detected in all parts of the water body. Two simple methods are used to measure the potential voltage or 'amount' of electricity in the water; both measure the difference in voltage between two points at differing distances from the source (Figure 4.5).

An illustration of the two methods of describing in-water voltage can be given by considering the set-up shown in Figure 4.5.

When contacts 'a' and 'b' of probe 1 are touching electrode 1, no voltage will be measured (both contacts are at the same potential as the electrode). As contact 'b' moves through the water towards electrode 2, the voltage (measured relative to electrode 1) will increase. If the shape and orientation in the water of electrode 1 and 2 are the same, at the halfway point the voltage will equal X volts ÷ 2. The voltage will reach a maximum (X volts) at electrode 2. The shape of the graph of the readings would look as shown in Figure 4.6.

Whilst probe 1 in Figure 4.5 measures relative voltage across a variable distance, probe 2 measures the voltage across a fixed distance and gives the gradient (E) of the voltage in volts per centimetre.

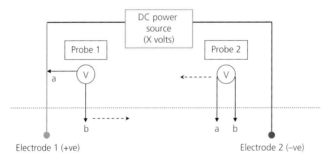

Figure 4.5 Measuring voltage in water. V = voltmeter.

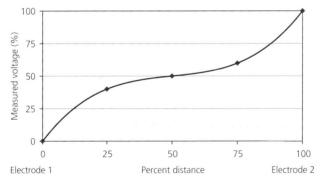

Figure 4.6 Voltage profile obtained from probe 1.

Voltage gradient can be measured (as for probe 2) in the water by use of a 'penny probe' (Figure 4.7) connected to either a digital voltmeter (DVM) or oscilloscope. The probe was so named by W.G. Hartley because it was practice for the end contacts to be made from copper pennies (W.G. Hartley, personal communication). By rotating the probe, the maximum and minimum values for the voltage gradient can be found for any position, and thus the field pattern can be plotted for any electrode or voltage combination. Care should be taken if using such a probe that no contact is made directly to the electrodes, and adequate insulation is used in the construction materials due to the high voltages that may be encountered.

As the probe moves between electrodes 1 and 2, the distance between contacts 'a' and 'b' is kept constant. Once outside a set distance from an electrode (10 to 20 radii for ring electrodes), the electric field, although still being present, should theoretically be so low as to have no effect on organisms within it. Knowledge of the voltage gradient profiles for different electrode arrays gives a basis for comparing the electric fields for different electrode configurations. Readings taken with probe 2 would look as shown in Figure 4.8.

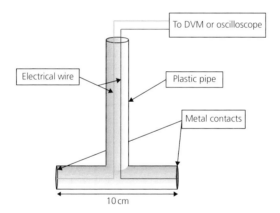

Figure 4.7 A simple probe for measuring voltage gradient.

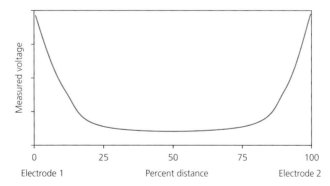

Figure 4.8 Voltage gradient profile obtained from probe 2.

The graphs shown here indicate the effectiveness with which a particular electrode can project power into the water. The symmetry of the 'U' and 'S'-shaped curves in Figures 4.6 and 4.8, respectively, is due to both electrodes having the same shape and orientation in the water. Unequal electrode resistance (discussed further in this chapter) will result in skewed gradients. The shapes of the 'U' and 'S'-shaped curves are important in electrode design. Poorly designed electrodes will not be able to project voltage well and will have an abruptly curving 'S' or a steep-sided 'U'. A steep-sided 'U' also denotes high voltage gradients that could be dangerous to fish. Well-designed electrodes, however, will propagate energy better, will have a shallower curve to the 'S' and 'U' and will not exhibit dangerous voltage gradients.

It is often considered that, for individual fish, the amount of stimulation is dependent upon the electrical potential between its head and its tail, and many researchers have noted the 'body voltage' which refers to this (Figure 4.9).

Results in the literature regarding the voltage gradients needed to incapacitate fish are extremely varied and often contradictory. In many cases, the waveform used has also not been sufficiently described, or key parameters (e.g. electrode size or water conductivity) have also not been noted.

In reality, however, fish are rarely so conveniently aligned, and stimulation is based upon the summation of the various voltage and power gradients encountered by the fish from a multitude of directions. Fish closest to the electrodes will, however, be subjected to higher gradients with consequent health and injury implications (Figure 4.9).

However, as a general rule, when using pDC in moderate water conductivity (~250 to 1000 μS.cm^{-1}) a voltage gradient of between 0.1 and 1.0 V.cm^{-1} is considered suitable for fishing, although there will be some variation with water conductivity, fish size and waveform used. The beginning of the forced swimming reaction occurs at around 0.1 V.cm^{-1}, and the onset of tetanus occurs at around 1.0 V.cm^{-1}. For DC, higher gradients are required due to the lower effect per unit volt for this waveform. The voltage required at the anode to produce this gradient at a particular distance is discussed in Chapter 6 on electrodes.

For pDC: attraction threshold = 0.1–0.2 V.cm

tetanising threshold = 0.5–0.6 V.cm

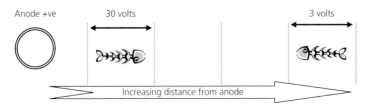

Figure 4.9 Diagram showing changing voltage and effect with distance from anode.

For DC: attraction threshold = 0.2–0.3 V.cm

                           tetanising threshold = >1.0 V.cm.

Voltage gradient profiles that have been measured for different diameters of ring electrodes (Beaumont *et al.* 2005) are shown in Figures 4.10. Figure 4.10a shows the higher gradients near the anode associated with small anode size, and Figure 4.10b (with log. *y*-axis) shows the increase in effective range of the larger anode sizes. The values shown are for 10 mm thickness rings energised at 200 V with a 750 mm braid cathode positioned 10 m away in the direction of measurement: the electrode is positioned at 0 cm. Note that some, but not all, of this difference will be due to changing anode resistance and Kirchoff's Law (Section 4.6.2).

Cuinat (1967) considered that the voltage propagation characteristics of a ring-shaped anode could be described by the propagation characteristics of hemispheres. From this relationship, he was able to calculate the voltage required to produce a particular voltage gradient at any distance from any sized anode. Cuinat (1967), however, refers to 'ambient voltage of anode' when actually he is referring to the total circuit voltage (Vt). This has important implications for the results due to Kirchoff's Law (see Section 4.6.2) and the change in voltage propagated from the anode when the anode–cathode resistance ratio is changed. Probably as a result of this, values calculated using data derived from Cuinat (1967) are in poor agreement with the data obtained from the measured and calculated values in Beaumont *et al.* (2005).

Beaumont *et al.* (2005) measured voltage gradients from a range of diameter ring anodes, corrected the readings for Kirchoff's Law (Section 4.6.2) and

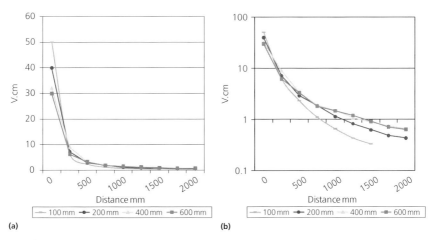

Figure 4.10 Voltage gradients measured from three different diameter ring (torus) anodes. Electrode is positioned at 0 cm. Circuit voltage was 200 V. (a) Normal *y*-axis; (b) log. *y*-axis to show the increase in low-gradient ranges.

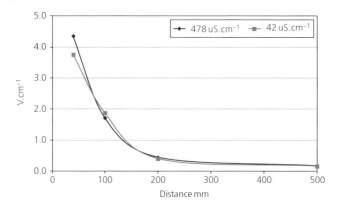

Figure 4.11 Voltage gradient of the same diameter ring at $478\,\mu S.cm^{-1}$ and $42\,\mu S.cm^{-1}$ specific conductivity and the same energising voltage (from Beaumont *et al.* 2002).

modelled the results. This enabled distances to specified voltage gradients to be calculated for a range of anode sizes and energising voltages. Four specified values of voltage gradient (0.1, 0.2, 0.5 and 1.0 $V.cm^{-1}$) from the ring anodes were modelled, from the empirically measured data, for each ring size.

Comparison of the model output with measured values indicated a precision well within 10% of measured values. Values were also well within the error that could be expected in the gradient threshold distance from causes such as variable bed conductivity (Scholten 2003) and so on. The equations for predicting the distance to specific voltage gradients have been incorporated into a spreadsheet (ElectroCalc). This calculates and predicts the voltage gradient parameters for a variable range of anode and cathode configurations: it is available from the author.

The voltage gradient for any given electrode configuration is constant for any water conductivity, provided anode voltage is kept constant (Figure 4.11). However, the gradient required to elicit a response from the fish will vary with different waveforms (AC, DC and pDC) and also water–fish conductivity ratios (Section 4.8.1).

4.3 Voltage waveforms

The voltage waveforms of electricity can be divided into two types:
1 **Alternating waveforms** (AC), characterised by the waveform going from a positive maximum to a negative minimum and continually reversing polarity (positive, then negative).
2 **Unipolar or direct waveforms** characterised by movement of electrons in one direction only. This waveform can be further sub-categorised into **continuous direct current** (DC) and **pulsed direct current** (pDC).

4.3.1 Alternating current

This waveform is the same as that generally used throughout the world for domestic electrical supply. The voltage direction reverses many times (50–60) a second; thus, there is not any polarity to the voltage (one electrode being successively positive and negative many times a second). AC may be single phase or multi-phase (a series of AC waveforms, usually three, offset in time from one another). Figure 4.12 shows the single-phase form of a voltage waveform.

This waveform has the advantage of being produced easily from small generators and suffers little variation in effectiveness due to physical parameters of the stream (streambed conductivity, weed beds etc.). Power loss from converting the AC current to DC (Power Factor – see Section 4.5.1) is also not an issue, meaning that smaller generators can be used. The voltage gradient required to provoke a reaction is also quite small.

When fish encounter an AC field, they experience:

• **Oscillotaxis**: The fish are attracted to the electrodes (but not to the same extent as with DC and pDC).
• **Transverse oscillotaxis**: The fish quickly take up a position across the current and parallel to the voltage gradient in order to minimise the voltage potential along their body.
• **Tetanus**: Once so aligned, the fish muscles are in strong contraction and the fish are rigid. Breathing is also often impaired by the fixation of the muscles controlling the mouth and opercular bones. The effect is more violent than with DC or pDC, and at high voltages muscular contractions may be so severe that the vertebrae and ribs are damaged. The recovery time can be significant.

The disadvantages of AC are that it has a very limited attraction effect to the electrodes; thus, fish are often immobilised beyond the reach of the net operator. Whilst the immobilisation zone of AC is large compared with those of DC and pDC, with no attraction zone the total effective 'capture' range is the smallest of the three waveforms (for similar voltage outputs). In addition, with little attraction to the electrode, fish are not drawn out of cover or deep areas to where they can be seen and caught.

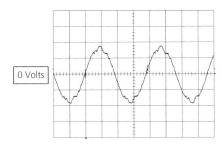

0 Volts

Figure 4.12 Single-phase AC voltage waveform.

In addition, the effect of AC is to tetanise fish with their muscles in a cramped state. This tetanus quickly restricts the fish's ability to breathe and renders them unconscious. If not removed quickly from the field, death may occur quite soon from asphyxia. Delayed mortality may also occur due to acidosis resulting from the oxygen debt generated by the contracted muscles. Kolz (1989) found that even when applying the same power to the fish, fish immobilised with AC took longer to recover than fish immobilised with pDC. Smolian (1944) considered that AC was a 'paralyser of fish' and because of this warned against using voltages higher than 150 V.

The detrimental impacts of AC waveforms are such that their use has been precluded from the European standard for sampling fish with electricity (Anon. 2003a). Snyder (2003) also recommends against its use for fish surveys in the United States, unless fish are to be killed and thus injury or mortality to uncaptured fish is not a concern. Its use for general surveys is not recommended. The waveform may have some use, however, for powering some pre-positioned arrays (discussed further in this chapter) due to the minimal attraction of the electrodes.

AC waveforms **can** be used as a primary power source (from a generator) to the control boxes that output DC and pDC waveforms.

4.3.2 Direct current

This is the simplest waveform used and technically is not a true 'wave' but a constant voltage applied over time (Figure 4.13). The electrical charge flows only in one direction (i.e. from negative (cathode) to positive (anode)).

Direct current was the first type of electrical waveform to be used for electric fishing (Baggs 1863), because it is the type that is produced from a galvanic cell (battery). Immobilising fish with DC needs a considerable amount of power (particularly at high water conductivities) and thus needs large generators, or if using batteries it will quickly exhaust them. Generators designed to produce DC current are heavier, more expensive, less reliable in voltage control and generally less reliable than AC generators with comparable power rating. For these reasons, DC power is usually produced by modifying power from an AC generator. A poor AC waveform, however, can affect the shape of the DC (and pDC)

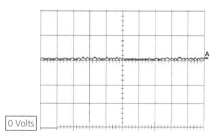

Figure 4.13 DC voltage waveform.

waveforms produced by some control boxes and can result in a noticeable ripple resulting from inefficient smoothing of the AC source current (Figure 4.13 shows a small ripple from this effect). Recent work on DC waveforms seems to show that this ripple, even if small (~2%), can have a marked effect on the fish. However, modern electronics should give a good, smooth DC waveform from an AC generator.

As the two electrodes (negative charge (cathode) and positive charge (anode)) produce differing physiological responses, the fish reaction will vary depending upon which electrode it is facing. The anode produces a highly attractive effect upon the fish within the capture zone. Fish so attracted will swim towards the anode along the current lines (see Figure 4.2) and so will have a curved trajectory towards the anode (Bohlin 1989). In contrast, the cathode creates a strongly repelling field, and fish are incapable of crossing, or at best extremely reluctant to face or cross, a cathodic field. This behaviour is used to great effect in the construction of electric barriers for excluding or diverting fish (see Chapter 10). When electric fishing in field situations, the cathode field should ideally be very diffuse (due to the large surface area of the electrode and thus lower electrical resistance) and should not overly influence the fish. Reactions to the anodic/positive DC field can be broadly categorised into five basic phases.

- **Alignment**: With initial electrical stimulation, the fish align themselves with the direction of the electrical current. If initially transverse to the anode, fish undergo anodic curvature that turns the fish's head towards the anode.
- **Galvanotaxis**: Once parallel with the current, fish start to swim towards the anode. This is caused by either electrical stimulation of the CNS, resulting in 'voluntary' swimming, and/or direct action of the electricity on the fish muscles.
- **Galvanonarcosis or narcosis**: When fish get close enough to the anode to experience a sufficient voltage or power gradient, their ability to swim is impaired. In this state, their muscles are relaxed.
- **Pseudo-forced swimming**: As the fish gets even closer to the anode, a zone where the fish begins again to swim towards the anode occurs. This swimming is caused by direct excitement of the fish muscles by the electric field and is not under the control of the CNS.
- **Tetanus**: At high-DC voltages, the muscles go from a relaxed state into spasm. This can result in impaired ability to breathe and possible skeletal damage.

Unless held under conditions of tetanus, when the electricity is switched off or the fish are removed from the electric field, they should recover instantly.

DC has a far greater attractive effect than other waveforms (AC and pDC), but it is less efficient as a stimulator and thus will not narcotise/tetanise the fish so readily. Thus the total capture zone of DC (attraction zone plus immobilisation zone) will be smaller than for pDC (for a fixed circuit voltage). The reason for this is because threshold values required to elicit responses are higher for DC than for AC and pDC. As it also shows great variation in effectiveness for slight variations in the physical factors that affect it (e.g. substrate conductivity), any such factors are likely to substantially reduce the effectiveness of the process. Kolz (1989) found that the DC 'stun' threshold (probably a reference to the tetanising threshold) was ~60% higher for DC than for either AC or pDC. The attraction threshold, however, was only 36% of that required to 'stun' with AC or pDC. The response of individual fish can also be somewhat variable to DC fields (Haskell *et al.* 1954). In general terms and for moderate water conductivity, DC voltage gradients of greater than 1.0 V.cm^{-1} are needed to create an incapacitating intensity and 0.2–0.3 V.cm^{-1} to create an attracting intensity. Overall, the narcotising voltage gradient for DC is often around twice that required for other waveforms. Dolan and Miranda (2003) found DC power thresholds required to immobilise fish with a volume of 50 cm^3 were twice that of 110 Hz pDC. A consequence of this is that DC may be less efficient (for a given voltage gradient) overall compared with AC or pDC. However, when fish do experience a DC intensity sufficient to immobilise them, they are in a relaxed state (narcosis rather than tetanus) and are thus not so likely to suffer injury.

Because the electrical field pattern around the anode is constantly changing, as the within-river physical configuration changes, it makes it difficult to standardise outputs for DC between sites.

4.3.3 Pulsed direct current

This waveform consists of pulses of DC electricity, with the number of pulses per second being measured in Hertz (Hz). In many ways, it acts like a hybrid between AC and DC. Like DC, it is unidirectional (i.e. it has no negative component), but it is not uniform. It has a low power demand (like AC) but is less affected by physical variations in stream topography (unlike DC). Voltage gradients required to elicit a response are also substantially lower than those for DC. The magnitude of the affect is also affected by the frequency of the pulses, implying that the cause of the effect on the fish may be different to that from DC.

The earliest pDC waveforms were created by interrupting a DC voltage (Leduc's current) and creating a 'square-wave' pattern of pulses. However, in the late 1940s, advances in electronics gave the ability to convert the negative component of an AC waveform to positive (full-wave rectification) or cut out the negative component (half-wave rectification) and thus create a pDC waveform (Figure 4.14a and 4.14b). From a standard 50 Hz AC generator, this gave either 100 Hz pDC (full-wave rectification) or 50 Hz pDC (half-wave rectification). The advantage of this method of creating a pDC waveform is the relative light weight and easy accessibility of AC generators compared with DC ones.

Further research demonstrated that a steep leading edge to the waveform provided the maximum physiological effect on the fish. Sternin *et al.* (1976) found that a sharp rise to the pulse was more 'effective' than a slow one but that anodic attraction was less. Vibert (1967) reported that early papers on electric fishing considered that the optimal pulse shape for electric fishing was a steep

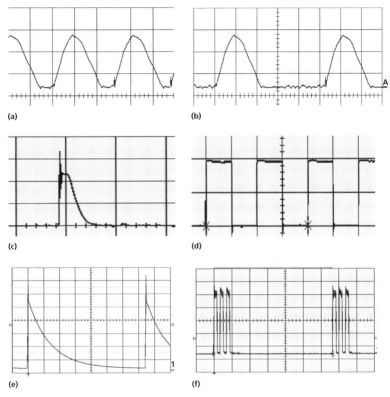

Figure 4.14 Examples of a range of pDC waveform types: note the 'spike' on the leading edge of waveform c.

increase and a slow decrease (e.g. Figure 4.14c). Novotny and Priegel (1974) also considered that there was evidence to suggest that the fast rise and slow decay of a quarter sine wave (Figure 4.14c) were advantageous for electric fishing. Further research, however (Lamarque 1967, Sharber & Carothers 1988), indicated that the efficiency of this waveform was due to its tetanising power, and thus such a waveform was the most damaging of the pDC waveforms. Due to these findings, quarter sine waveforms have been banned in some states in the United States. It is not clear whether tests have been carried out at lower voltage or power settings, and it may be that the tetanising threshold is just lower for this waveform.

Another waveform that is often used is capacitance discharge, also known as exponential pulse (Figure 4.14e). As its name suggests, this waveform is usually produced by charging a capacitor, which is then discharged through the electrodes. Because this discharge is of short duration, this pattern has the advantage that high peak voltages are available for fishing whilst, because pulse width is small, mean loading on the power source is small (i.e. RMS voltage and power are low). Problems exist with this waveform, however, as discharge duration is normally determined by the electrical conductivity of the water and thus cannot be easily controlled and allow equivalent power settings to be used at different sites. Stewart (1990) found that for pulse widths below 1 ms, exponential pulses required 2–5 times the voltage gradient of square-wave pDC to achieve the same degree of (excised) muscle reaction (tension). As with the quarter sine waveform, some evidence suggests that this waveform can cause high injury rates to fish in moderate and high-conductivity water (Lamarque 1967). Sharber and Carothers (1988), however, found that injury rates for exponential pulse waveforms were no worse than for square waveforms.

Many of the newer designs of pulse boxes use square waveforms (Figure 4.14d). This waveform combines the advantage of good physiological effect with the ability to control and replicate pulse duration and frequency, thus allowing standardised output parameters (including power) to be used.

With the recent advances in electronics, it is now possible to produce a wide variety of waveforms from the basic AC generator or DC battery supply: the pulse box on the Smith-Root Inc. LR-24 backpack, for example, can produce over 250 different waveform variations. Modern electric fishing control boxes can also have the facility to produce a variety of nonstandard waveforms. However, the principles behind these (e.g. decreasing pulse interval, and high to low-frequency variation) are probably not valid in real-life situations where fish may be appearing from all directions simultaneously. Until evidence shows some benefit from these novel waveforms' use, they are best avoided. An exception to this is the 'Gated Burst' waveform or modulated pulse (Sternin *et al.* 1976) which is shown in Figure 4.14f and Figure 4.15. This waveform is basically a series of high-frequency pulses repeated in a lower frequency pattern. The number of pulses, the frequency and pulse width of the high-frequency pulses in the burst as well as the frequency of the repetition of the bursts can all be programmed (Figure 4.15). The waveform appears to have low injury rates and also very low power demand. Thus, some advantages may be obtained from this waveform in

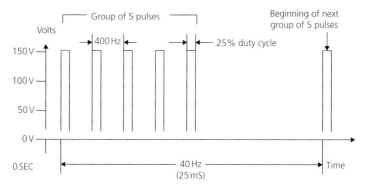

Figure 4.15 Detailed diagram of a Gated Burst waveform showing the parameters that are possible to set (with permission from Smith-Root Inc.).

terms of reducing fish injury, and for conserving power when it is limited (e.g. very conductive water or when using battery-powered equipment; the author has used this pulse type to fish successfully in 3200 µS.cm^{-1} water using the Smith-Root LR-24 battery-powered backpack).

Several studies have been published assessing the physiological effect of pDC electrical waveforms on fish (Sharber & Carothers 1988, McMichael 1993, Hollander & Carline 1994, Dalbey *et al.* 1996, etc.). Sharber and Carothers' (1988) study gives guidance regarding the 'best' pDC waveforms to use; square-wave and exponential pulse each give 44% injury rates, and sine-wave a 67% injury rate. From the author's personal experience, however, these seem very high; and in the large-scale juvenile Atlantic salmon monitoring program with which the author has been involved for the past 15 years (>10,000 individuals caught annually), injury rates of less than 2% were achieved (using either square-wave or half-wave rectified pDC).

Few studies accurately quantify the electrical characteristics of the waveforms being used. It is rare, for example, to either show oscilloscope traces or note that the traces have been seen and are what they purport to be. It has also been shown that the description of the waveforms can be wrongly described (e.g. Hill & Willis 1994, Dalbey *et al.* 1996). The problem is ably demonstrated by tests carried out on pulse boxes where waveforms were affected by the generator voltage output characteristics (Beaumont *et al.* 2002). Another problem with many of the studies reported is that some of the commercially available pulsing boxes have large transient voltage spikes superimposed on the specified waveform (Jesien & Horcutt 1990, Beaumont *et al.* 1997 and personal observation). An example of this is shown in Figure 4.14c. Inadequate recording of electrical details in many of the studies on electric fishing (e.g. no oscillograph traces) makes it difficult to identify the studies where these transients may be present. Even where voltage levels are recorded, if these are presented for RMS voltage levels instead of peak voltage levels, the effect of the transients will not be adequately recorded. In studies using equipment producing transient spikes, if peak voltages are back-extrapolated from mean voltages (Thompson *et al.* 1997), considerable errors may occur. The effect of these transients is largely ignored in discussions on waveform and electric fishing effect. However, Haskell *et al.* (1954) noted that the response of fish (to an electric field) was not improved by waveforms with a high initial peak. Jesien and Horcutt (1990) found that the equipment they were using produced a spike that increased with increasing water conductivity. Information is limited, however, and further research needs to be carried out on the impact and importance of transient voltage spikes on fish. This lack of definitive knowledge of the shape of the waveforms used in the majority of the research makes much of the findings difficult to apply and extrapolate to other studies.

The behaviour of fish to pDC is somewhere between that of AC and DC. As with DC, the fish react differently to the anode and cathode fields, and thus their reaction will vary depending on which electrode they are facing. There is some debate among researchers as to whether pDC produces true galvanotaxis and whether narcosis or tetanus causes immobilisation. In general terms, however, a fish's reaction to a pDC field can be summarised as follows.

- **Electrotaxis**: There is good attracting power, but this is due to the electrical effect on the fishes' muscles (the muscles contracting with each pulse of electricity and thus accentuating the swimming motion) and not, as may be the case with DC, by electrical effect on the spinal nerves. This vigorous effect upon the fish can also increase injury rates.
- **Tetanus and narcosis**: Like DC, the fish are immobilised near the anode but at a much lower voltage gradient; as tetanus may be involved, the fish need to be removed from this zone quickly.

As stated, voltage gradients required to elicit a response are lower for pDC than for DC. Few data exist, though, detailing the gradients that are required. Edwards and Higgins (1973) noted that to immobilise bluegill, a pDC gradient of 0.66 $V.cm^{-1}$ was required, compared with a 1.66 $V.cm^{-1}$ for DC. Davidson (1984) showed that the pDC voltage gradient required for immobilisation differed between species and varied with pulse frequency; average values were about 0.4 $V.cm^{-1}$ and were constant above 50 Hz.

Experience has shown that minor changes in physical parameters within the stream have little impact upon the efficiency of a suitably set-up pDC system, thus making the efficiency of the waveform more uniform than DC both within and between sites. In addition, pDC waveforms have the additional advantage that it is possible to alter the applied power to the water by both increasing the pulse frequency (with constant (milliseconds) pulse width) and varying the pulse width (see Sections 4.2.3.1 and 4.2.3.2). Research has shown, however, that pDC is more stressful and causes more injuries than DC (Lamarque 1967, 1990, McMichael 1993, Dalbey *et al.* 1996). Fish immobilised with pDC can also have greater recovery times than fish immobilised with DC (Mitton & McDonald 1994).

Figure 4.16 gives a summary of the key characteristics of the main waveform types.

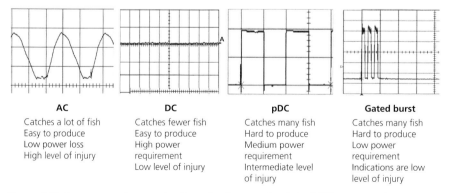

AC	DC	pDC	Gated burst
Catches a lot of fish	Catches fewer fish	Catches many fish	Catches many fish
Easy to produce	Easy to produce	Hard to produce	Hard to produce
Low power loss	High power	Medium power	Low power
High level of injury	requirement	requirement	requirement
	Low level of injury	Intermediate level	Indications are low
		of injury	level of injury

Figure 4.16 A summary of properties (for a fixed power density) of different waveforms.

In Canada (Alberta) and the United States (Montana), 60 Hz quarter sine, capacitor discharge and AC waveforms have all been banned due to concerns over fish injury with these waveforms. However, the research upon which this policy is based is not explained in the policy document. Overall, it would seem that damage can be caused by all pDC waveforms, and little 'improvement' over the original full-wave and half-wave rectified shapes has taken place. Square waveforms, whilst still having rapid changes in voltage, do have the advantage of being able to have their output parameters (pulse width and voltage) more accurately controlled and quantified than many of the other waveform types.

4.3.3.1 Pulse frequency

A single cycle or pulse of AC or pDC voltage is defined as the complete sequence intervening between two successive corresponding points in a regularly recurring sequence of potential variations in a periodic voltage waveform – in simple terms, from the start of one cycle/pulse to the start of the next cycle/pulse. The frequency of the pulses or cycles is measured in pulses per second, and the unit of measurement is the Hertz (Hz). In Figure 4.17, the blue line is the pulsed (pDC) waveform, and the green arrow denotes one cycle of the waveform (the start of one pulse to the start of the next). If the red arrow is 1 s long, then the frequency will be 2 Hz as there are two complete cycles in the time.

Until recent advances in electronics and pulse box design, only two pulse frequencies (50 and 100 Hz) were commonly used for pDC electric fishing. The principal reason for this is that, historically, the source of the electricity was a commercial generator (producing 50 Hz AC), and the pulse box either full wave rectified the AC (producing 100 Hz pDC) or half wave rectified the AC (producing 50 Hz pDC). Modern equipment, however, enables a wide variety of pulse frequencies to be used, and considerable experimentation has taken place regarding the most efficient pulse rates to capture different species. Stewart (1990), working on excised fish muscle, considered that muscles of differing species of fish have a maximum contraction rate; for plaice (*Pleuronectes platessa*), this was 20 Hz. Above this, rate muscles become tetanised and the fish is immobilised. Justus (1994) and Corcoran (1979) found that optimal frequency varied

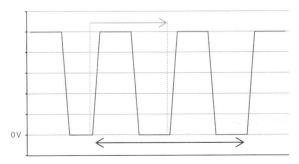

Figure 4.17 Schematic diagram of definition of a pulse frequency.

Table 4.1 Optimal tetanising frequencies for different fish species.

Species	Optimal frequency (Hz)
Minnow	90
Trout	80
Carp	50
Eel	20

Halsband (1967).

between similar catfish species. Novotny and Priegel (1974) stated that some species selectivity when electric fishing is possible by varying the pulse frequency. Halsband (1967) stated that the frequencies shown in Table 4.1 are optimal for those species (although the term 'optimal' is not explained and could mean tetanising values). For example, in an experiment carried out by Lamarque and reported in Vibert (1967), the 'optimal' frequency (of a square-wave 33% duty cycle waveform) for creating anodic taxis in a 20 cm trout (at 18 °C) was 100 Hz. However, this frequency was not recommended, as the tetanising power of this frequency was also optimal. Lower frequencies of 4 to 10 Hz were recommended. This raises an important concept that perhaps researchers should not be looking for 'optimal' or 'efficient' frequencies (and certainly not for tetanising effects) but for 'benign' ones (Hickley 1985). Bird and Cowx (1993) also found that fish were immobilised more effectively at 100 Hz, but long recovery periods imply that the fish were being tetanised.

Lamprey larvae (ammocetes), one of the species identified in the questionnaire in Beaumont *et al.* (2002) as difficult to capture, have been successfully caught by using slow (3 Hz) pulse rates (Pajos & Weise 1994) to attract the ammocetes, and then switching to higher frequency (40 Hz) to tetanise them for capture (Weisser & Klar 1990).

Much of the research published regarding the effects of different pulse frequencies is far from clear. Differences exist between pulse shapes, voltage gradients, pulse widths, species and so on used, making it difficult to isolate individual components' effects. Plus, there is the problem that transient voltage spikes may not have been noticed or recorded. However, the research does support the proposal that as frequency increases above 15 Hz, injury levels increase (Snyder 1995, Sharber *et al.* 1994, McMichael 1993, Cooke *et al.* 1998 etc.). Figure 4.18 summarises these findings.

Reynolds and Holliman (2002) consider that 'if injury is to be significantly reduced', frequency should be reduced to 15–30 Hz. The exception to this is the use of the gated bursts where high-frequency bursts are created at moderate-frequency intervals.

The cause for these injuries has yet to be fully understood, but high-frequency voltage passes more easily into animal tissue than low-frequency voltage

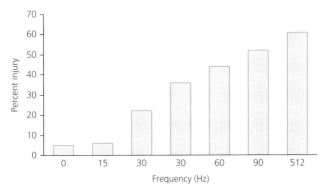

Figure 4.18 Percentage injuries for different frequencies of square wave pDC (0 Hz = DC) (collated from sources quoted in text).

(Maletzky 1981, quoted in Sharber *et al.* 1995). Sternin *et al.* (1976) found that increasing the frequency of the waveform resulted in apparent fish conductivity also increasing (due to a decrease in the capacitance of the fish muscle cell membrane), thus implying that at higher frequencies, more current would pass into the fish. Collins *et al.* (1954) considered that the danger point was when the current 'switched on', and this is supported by Sharber and Black (1999) who consider that the underlying principle behind electric fishing effects is induced epilepsy. If correct, this could also explain the occurrence of injury even with DC fields, the injury occurring when the DC is switched on (Sternin *et al.* 1976, Sharber *et al.* 1995). Supporters of Power Transfer Theory (Kolz 1989, Kolz & Reynolds 1989, 1990) consider that the higher frequency delivers more power into the fish and thus increases the likelihood of injury (see Section 4.7.1). This would also help explain the increased injury rates of higher frequency waveforms. Whichever theory is correct, the fact remains that more fish injuries are caused by high frequencies of pDC; and therefore, when fishing, frequency should be kept to a minimum (below 60 Hz).

The effective range of an anode will also be affected by the frequency used. However, Davidson (1984) found that in tank-based trials on roach, perch, pike and eel, the distance of immobilisation did not always increase with increasing frequency. Results using 10% pulse width are shown in Figure 4.19 (note that no immobilisation occurred in rainbow trout at 10 Hz and 10% pulse width). Davidson (1984) ascribed these differences to the presence of optimal frequencies where reaction is greater for the different species (as described by Halsband 1967 in this section).

For a fixed (millisecond) pulse width, altering the pulse frequency will have an effect on the average (RMS) power used by the fishing set-up (Figure 4.20). Low pulse frequency therefore is not only beneficial to fish welfare but also will result in power savings. Note that peak power will remain constant whatever the pulse frequency; this is important when equipment output is limited by peak power.

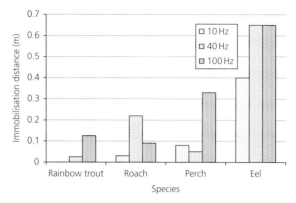

Figure 4.19 Immobilisation distance (m) at differing frequencies for four fish species.
Note: There was no response at 10 Hz, 10% duty cycle for rainbow trout.
Adapted from Davidson (1984).

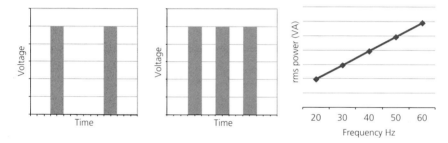

Figure 4.20 The effect of increasing pulse frequency on applied power.

4.3.3.2 Pulse width

Pulse width is, as the term suggests, the duration of the individual pulses of electricity. There are two ways of expressing this factor.

- **Pulse width**, expressed in milliseconds (ms) duration that the current is flowing
- **Duty cycle**, expressed as the percentage (%) time within one cycle that the current is flowing.

This can lead to some confusion: for example, a 50% duty cycle at 50 Hz (10 ms pulse width) will have a different pulse width from a 50% duty cycle at 100 Hz (5 ms pulse width) (Figure 4.21).

Duty cycle can be calculated from pulse width (in seconds) if the frequency is known:

$$\text{Duty cycle} = \left(\text{Pulse width}/\left(1/\text{frequency}\right)\right) \times 100 \qquad (4.1)$$

Increasing pulse width will increase the mean power required for the fishing set-up (Figure 4.22), but note that the peak power will remain the same. As power requirements for electric fishing are most easily initially calculated for a

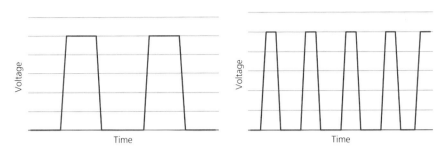

Figure 4.21 Diagrammatic representation of 50% duty cycle at different frequencies, showing different pulse width duration; time-base is the same in both diagrams.

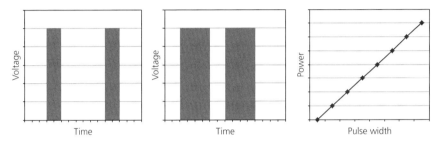

Figure 4.22 The effect of increasing pulse width on power demand.

DC waveform, expressing pulse width as a duty cycle makes calculation of the power demand for square waveform pulses (the most common waveform in use) easier as, for example, a 10% duty cycle will use 10% of the power needed by DC and half the power of a 20% duty cycle, and so on.

Whilst some research has been carried out regarding the effects of increasing pulse width (Halsband 1967, Daniulite & Prits 1965 quoted in Vibert 1967, Sternin *et al.* 1976, Davidson 1984, Stewart 1990, Bird & Cowx 1993, Miranda & Dolan 2004), it is often contradictory. The obvious effect of increasing pulse width is to increase the mean power (but not the peak power) transmitted into the fish. Several early researchers have found that once a threshold in pulse width has been reached, increasing it above that has little further effect, and the energy is 'wasted'. Halsband (1967) referred to this time as the 'effective duration', whilst Sternin *et al.* (1976) referred to it as the 'Chronaxy time', which is twice the time taken for the effect to manifest itself. Most research in freshwater reports that pulse widths of between 5 µs and 5 ms are adequate for fish capture in a wide range of conditions. Work by Daniulite and Prits (1965) on herring (in seawater) also found that pulse widths of between 0.2 and 0.56 ms were adequate for creating anodic reaction. They also noted that higher pulse widths were required when the pulse frequency was lowest (<25 Hz), and Halsband (1967) similarly stated that if pulse width is reduced (below the effective duration), higher voltages are required. Working on excised fish muscle, Stewart (1990) found that as duration (of pulse) is increased, smaller (lower voltage) pulses produce the same tension in the muscle. All of these findings could relate to Kolz's Power Transfer Theory (PTT), where fish require a minimum power to elicit a response. However, Kolz (1989) uses peak values of power transfer, and therefore (according to the PTT) adjusting pulse width (which affects mean power) should not change the fish's reaction. Miranda and Dolan (2004), investigating ways to induce narcosis rather than (damaging) tetanus in fish, found that duty cycles of 10–50% required the least power to immobilise fish. This range of duty cycles also resulted in the highest margin of difference between the power needed to narcotise, rather than tetanise, the fish. Peak power required (for immobilisation) increased gradually above the 50% and sharply below the 10% thresholds.

Davidson's (1984) findings regarding the range of immobilisation for differing pulse widths are shown in Figure 4.23. Immobilisation distance increased with increasing pulse width. However, attraction distance was not so well correlated. At 100 Hz only two species, at 40 Hz one species and at 10 Hz one species showed slight increases in attraction distance. Work by Bird and Cowx (1993) also revealed poor correlation between the voltage gradients required to elicit a response from the fish and pulse width.

Whilst adjusting either pulse width at constant voltage, or voltage at constant pulse width, is a valid way of adjusting the **mean** power, there will be differences in effect depending upon which method is used. In the case of constant voltage with variable pulse width, the **peak** (instantaneous) power would be the same for all settings of pulse width. In the case of constant pulse width with variable voltage, the **peak** (instantaneous) power would change (as a function of the square of the voltage). The effects of these different methods may be the cause of some of the variation in results observed by various authors.

Beaumont *et al.* (2000) examined both efficiency of capture and stress response (as measured by blood plasma cortisol levels) between a range of waveforms. No difference was found between the catch efficiency or stress between 6 ms (36% duty cycle) and 5 μs (3% duty cycle) pulse width square waves. Catch per unit

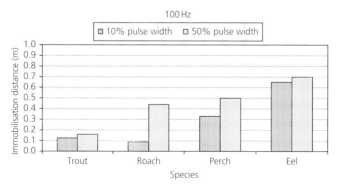

Figure 4.23 Difference in effective ranges for immobilisation between 50% and 10% pulse widths for four fish species at 100 Hz frequencies (adapted from Davidson 1984).

power of the 5 μs pulse width, however, was around nine times that for the 6 ms pulse width, indicating its efficiency in terms of power usage.

Notwithstanding the uncertainty regarding the cause of its effect, adjusting pulse width at a constant voltage is the usual method employed to increase power in high-conductivity waters. It should be noted, however, that the 'power' control on many old pulse boxes in use in the United Kingdom alters both the peak voltage and the pulse width of the waveform (Beaumont *et al.* 2002). For example, at the minimum setting, the Electracatch WFC4-20 produced a 2 ms pulse of 84 V_{pk}, and increasing the 'power' to maximum resulted in a 12 ms pulse of ~350V_{pk}. This voltage increase is the opposite of what operators should try to achieve in higher conductivity water, as it will increase the likelihood of injury to the fish.

4.4 Electrical current

Electrical current is the quantity of charge moving through the circuit per unit time (Coulombs per second). It is measured in amperes (often shortened to amps (A)), and the SI dimension symbol is 'I' (derived from the French term *intensité de courant*). Amps can be either measured using an ammeter or calculated from:

$$\text{Current}\,(\text{amps}) = \text{volts} \div \text{resistance} \tag{4.2}$$

From Equation (4.2), it can be seen that the current drawn by a particular electric fishing set-up and circuit voltage will vary depending on the water resistance (conductivity). On many older pulse boxes used in the United Kingdom, the only indication of electrical output is an ammeter, and it is important to note that the same ammeter readings at different sites could reflect very different in-water electrical voltage output properties.

For example, if fishing a low-conductivity stream (e.g. 100 μS.cm^{-1} using a 400 mm anode and 3 m braid cathode), the circuit equivalent resistance (see later for a full description of equivalent resistance) will be about 160 Ω. If circuit voltage is 300 V and waveform is DC, then current will be 200 ÷ 160 = 1.9 A. However, if fishing a high-conductivity stream (e.g. 800 μS.cm^{-1}) with the gear configuration described above, circuit equivalent resistance will be 20 Ω, so, 300 ÷ 20 = 15 A: in other words, approximately 10 times higher. Note that if the amperage in the high-conductivity site was too high for the pulse box capacity, then using pDC at a 25% duty cycle would drop the mean amps to around 3.75 A (25% of 15).

Current density (J) is the strength of the electrical current (per unit area cross-section) at a particular point in the water; it is measured in amps.cm^2.

$$J = E \times c \tag{4.3}$$

where J = current density; E = voltage gradient; and c = water conductivity.

Many researchers (e.g. Smolian 1944, Monan & Engstrom 1962, Kolz 1989, Kolz & Reynolds 1990, Sternin *et al.* 1976) consider that current density is the most significant factor in determining a fish's reaction to an electrical field. By integrating a measure of water conductivity into the parameter, it also enables standardisation of outputs in differing-conductivity water (see also Section 4.8.1.3 on PTT for a description of power density and output standardisation).

4.5 Power

Electrical power (symbol P) is the energy per unit time, or the rate that electrical energy is transferred. The unit of measurement of power is the watt (W) and equals a power transfer rate of 1 Joule per second. Power can be calculated from any of the following:

$$\text{Power} = V^2 \div R \tag{4.4}$$

$$= I^2 \times R \tag{4.5}$$

$$= I \times V \tag{4.6}$$

where V = voltage; I = amperage; and R = resistance.

As power is a function of the voltage and the resistance of the circuit, the power required to power any fishing system for any known water conductivity can be estimated if the electrode equivalent resistance (see later) and required voltage for the water conductivity are known.

From the equations above, where resistance is a divisor in the equation, it can be seen that, for a fixed voltage or current, in high resistance (low-conductivity) waters less power is needed to propagate an electric field in the water than in low-resistance (high-conductivity) waters.

For example, if fishing a circuit voltage of 200 V DC in a low-conductivity stream, giving a circuit equivalent resistance of 163Ω:

$$\text{Power} = 200^2/163 = \sim 250\,\text{W}.$$

For a high-conductivity stream with a circuit equivalent resistance of 20 Ω;

$$\text{Power} = 200^2/20 = 2000\,\text{W}.$$

For low-conductivity water, this apparent reduction is to some extent offset by the need to use a higher voltage to immobilise the fish in low-conductivity systems.

With very high-conductivity water, the available power is often the limiting factor with regard to the ability to electrify fish. Reducing the size of the electrodes can lessen power use in high-conductivity water. This method, however, can result in high field intensities being created near the electrodes and/or smaller field patterns from the electrodes.

Alternatively, using pDC and reducing the frequency (if using a fixed millisecond pulse width) or the duty cycle of the waveform can be used to reduce the power demand. For square-wave waveforms, the power required is directly proportional to the duty cycle of the waveform (i.e. a 25% duty cycle will use 25% of the power of a DC waveform). In practical field use, the actual power values will vary about the calculated value depending upon streambed conductivity variation and the proximity of the anode to the cathode (and, in twin-anode situations, the proximity of the anodes to each other): calculated values should therefore be regarded as a minimum.

Power input required to energise electric fishing gear using a standard 400 mm diameter anode and a 3000 mm braid cathode designs and DC waveform can be very high. If 3.0 kVA is the likely upper limit of generator size suitable for portable field use (based on estimated weight of 35 kg), and 250 V DC input is the likely minimum applied voltage required, a water conductivity of about $400\,\mu S.cm^{-1}$ is the maximum upper limit for DC electric fishing. Larger systems can be used, and French researchers fish water up to $1000\,\mu S.cm^{-1}$ using DC; however, the generator size is extreme, with the estimated weight of the generator being 90 kg! Boat-mounted systems can be more portable, but even these are limited by weight to about 7.5 kVA. In water of higher conductivity, pDC will need to be used if generator size is to remain truly portable.

The input power requirement for a 400 mm diameter anode (and a 3000 mm braid cathode) at different applied DC voltages is shown in Figure 4.24. The y-axis has been limited to 3.5 kVA as this is about the practical limit for a mobile (hand-carried) AC generator (see Section 4.7.1).

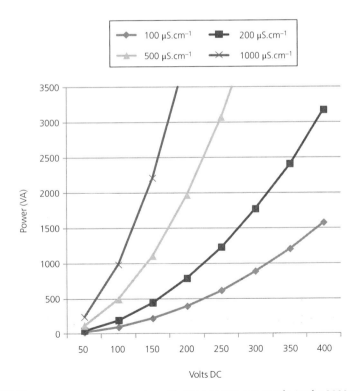

Figure 4.24 The power needed to energise with DC one 400 mm anode (and a 3000 mm braid cathode) at different water conductivities (from an AC generator).

4.5.1 Power factor

If using an AC generator to energise a DC (or pDC) circuit, inductance and capacitance issues may result in more power being required than circuit theory would predict. This difference between 'apparent power' (the power actually required) and the 'real power' (theoretical) is called the 'power factor' (PF). It means that a correction (that will increase the power demand) needs to be added when assessing the theoretical power required.

For standard generator-derived circuits, the power factor is always between 0 and 1; when it is 1, it is termed the 'unity power factor' and apparent power = real power. The inefficiency of the electric fishing control box will depend upon the circuitry within the unit. It is technically feasible to design circuitry with a unity power factor with little detriment to conversion efficiency, and modern units powered by inverter technology generators are approaching this. Compliance with electromagnetic compatibility (EMC) regulations for the pulse box, however, often result in higher power factors.

In practice, with currently available equipment, it may be anticipated that the power factor may be as low as 0.6, meaning that a generator of about 150% calculated power demand for the fishing circuit is required.

When real power is converted to apparent power by using a power factor correction, the unit of power changes from watts and becomes the volt.amp (VA): generators are commonly rated by their capacity in kVA (1 kVA = 1000 VA).

4.6 Resistance and resistivity

The opposition to an electric current passing through a circuit or object is called its 'resistance', and it is measured in ohms (Ω).

Resistivity is how strongly a particular substance opposes an electrical current. Resistivity takes into account the shape of the substance and is measured in ohm-metres (Ω.m). Both resistance and resistivity are affected by temperature, decreasing as the temperature rises.

Knowledge of the resistance of a circuit is important for calculating the current demand (amps) and the power needed to energise the circuit (watts/kVA). Where a circuit has more than one resistor in it, the total resistance of the circuit needs to be calculated. This total resistance is the combined resistance value of all the resistors in the circuit.

A simple single-anode electric fishing circuit is shown diagrammatically in Figure 4.25. The resistors Ra (anode) and Rc (cathode) are in series, and the resistance of the circuit is simply calculated by adding the two resistance values. If Ra = 20 Ω and Rc = 10 Ω, then the total circuit resistance will be 30 Ω (from Equation (4.7)).

$$R = Ra + Rc \qquad (4.7)$$

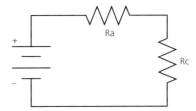

Figure 4.25 Circuit diagram of resistors in series.

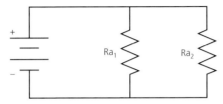

Figure 4.26 Circuit diagram of two resistors in parallel.

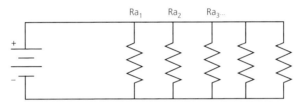

Figure 4.27 Circuit diagram of more than two resistors in parallel.

Where two anodes are used, the electrodes are wired in a parallel configuration (Figure 4.26), and the calculation of the resistance of the two anodes is more complex. If the anodes Ra_1 and Ra_2 are both 20 Ω, then, from Equation (4.8), circuit resistance is 10 Ω. As can be seen, unlike a series circuit where additional resistors increase the circuit resistance, in a parallel circuit extra resistors decrease the circuit resistance: for two resistors of equal value, the circuit resistance is about half the value of an individual single resistor of the same value.

$$R = \left(Ra_1 \times Ra_2\right)/\left(Ra_1 + Ra_2\right) \tag{4.8}$$

Where more than two anodes are wired in parallel (Figure 4.27), the equivalent resistance is calculated from:

$$R = \left[1/\left[\left(1/Ra_1\right)+\left(1/Ra_2\right)+\left(1/Ra_3\right)+\text{etc.}\right]\right] \tag{4.9}$$

For an electric fishing circuit, however, we also need to consider the cathode. So, for a twin-anode, single-cathode circuit (Figure 4.28), the equation would be

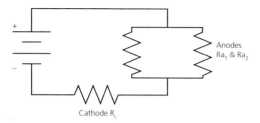

Figure 4.28 Circuit diagram of two anodes in parallel and one cathode in series.

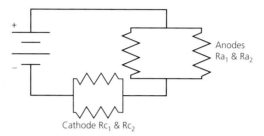

Figure 4.29 Circuit diagram of twin-anode and twin-cathode circuit.

the value of the anodes in parallel (shown within the [] brackets), plus the value of the cathode in series.

$$R = \left[(Ra_1 \times Ra_2)/(Ra_1 + Ra_2) \right] + Rc \qquad (4.10)$$

If Ra_1 and Ra_2 are both 20 Ω and Rc is 10 Ω, then, from Equation (4.10):

$$R = \left[(20 \times 20)/(20 + 20) \right] + 10$$
$$= \left[10 \right] + 10$$
$$= 20\,\Omega$$

For a twin-anode, twin-cathode circuit (Figure 4.29):

$$R = \left[(Ra_1 \times Ra_2)/(Ra_1 + Ra_2) \right] + \left[(Rc_1 \times Rc_2)/(Rc_1 + Rc_2) \right] \qquad (4.11)$$

For this twin-anode plus twin-cathode arrangement, if Ra_1 and Ra_2 are both 20 Ω and Rc_1 and Rc_2 are both 10 Ω, then, from Equation (4.11):

$$R = \left[(20 \times 20)/(20 + 20) \right] + \left[(10 \times 10)/(10 + 10) \right]$$
$$= \left[10 \right] + \left[15 \right]$$
$$= 15\,\Omega$$

The importance of the resistance ratio between the anode and the cathode is discussed in Section 4.6.2 on Kirchoff's Law.

4.6.1 Electrode resistance

In electric fishing, the electrodes themselves are made from a low-resistivity metal, usually aluminium, stainless steel or copper. They are linked to the pulse box connectors by means of copper cables and, in air, have a very low electrical resistance. Even with very long cables, the value of resistance from the connector to any point on the anode or cathode may be expected to be less than $0.25\ \Omega$. If the electrodes were to come into electrical contact, then the total circuit resistance between the two connectors would be the sum of the two individual electrode resistances (e.g. $2 \times 0.25\ \Omega = 0.5\ \Omega$).

However, when two electrodes are separated in water, the electrical resistance between the electrodes is a function of two components. One component is the sum of the two individual (in air) electrode resistances (e.g. $0.5\ \Omega$ in the example above). The other, which is a major component, is a combination of the resistance of the water, the spacing of the electrodes, and the dimensions and geometry of the electrodes. Orientation of the electrodes has only a minor impact upon their resistance provided that:

The spacing of the electrodes is large compared to the dimensions of the electrodes.

The volume of water is large compared with the spacing.

The electrodes are not close to boundaries such as the surface, bottom or sides of the river or lake.

Thus, the resistance of electrodes in water is equivalent to a higher value than that of the electrodes in air. This combination of resistances is termed the 'equivalent resistance' of the electrodes (R_{eq}). The equivalent resistance can be affected by insulating corrosion building up on the electrodes; for example, on aluminium electrodes, aluminium oxide (which is often used for insulating electric wires) can rapidly build up. It is important, therefore, that electrodes are kept clean.

Electrode resistance can either be calculated from theoretical principles (Cuinat 1967, Novotny & Priegel 1974) or empirically determined by field measurements. Some discrepancies will occur between the two methods of estimation, however, due to inherent differences between the theoretical and real world (Beaumont 2005).

Novotny and Priegel (1974) described a theoretical relationship to evaluate the equivalent resistance of electrodes and provide the graphs, and a graphical method to calculate it.

Empirical values of equivalent resistance for electrodes can be measured for a particular electrode by measuring the total circuit resistance using two physically identical electrodes. Equivalent resistance for each electrode will be half the total resistance measured. Once the value is known for a particular electrode, then the value for a second (different) electrode can be calculated by evaluating the total circuit resistance (from Ohm's Law) for two electrodes (one known and one unknown) and then subtracting the equivalent resistance of the known electrode, to give the value of equivalent resistance for the unknown electrode.

For example, if the resistance of electrode 1 is $20\,\Omega$ and the resistance of electrode 2 is unknown, then if the measured or calculated circuit resistance is $50\,\Omega$, electrode 2's equivalent resistance is $50-20 = 30\,\Omega$.

In practical electric fishing scenarios, electrodes are commonly used close to a boundary layer that affects the resistance characteristics. In the case of an anode, the adjacent boundary would usually be the surface of the water; in the case of a cathode, the adjacent boundary would usually be the bottom of the stream or water body. This can have the effect of increasing the value of the equivalent resistance of the electrode by up to twice the 'free water' value, and the value when measured near the boundary is the highest.

Kolz (1993) gave some empirical equivalent resistance values for different anode shapes, and Beaumont *et al.* (2002, 2005) gave calculated resistance values (from measured volts ÷ measured amps) for a range of sizes of ring anodes (Table 4.2) and different designs of cathode.

Cathode resistance was also measured in a similar manner and, for flat plate/mesh cathodes, found to increase in a cube relationship to area (Table 4.3). For a 25 mm width copper braid, there was a power law relationship with increasing length.

Table 4.2 Calculated equivalent resistance values for a range of ring anode sizes (ambient conductivity 350 μS.cm^{-1}), measured with electrode at surface of water (i.e. highest value).

Electrode size (mm)		Applied volts	Measured amps	Calculated resistance Ω
Diameter	Gauge			
100	6	51.0	0.4	127.5
200	6	50.9	0.5	101.8
400	6	51.0	0.7	72.9
600	6	50.8	0.7	72.6

From Beaumont *et al.* (2003).

Table 4.3 Cathode sizes evaluated and equivalent resistance (measured at 350 μS.cm^{-1} ambient conductivity) with cathode on the bottom of the river.

Cathode size (mm)	Cathode material	Braid thickness/ mesh size (mm)	Resistance (Ω)
750 × 25	Copper braid	3	48
1500 × 25	Copper braid	3	31
3000 × 25	Copper braid	3	20
250 × 250	Steel mesh	13	49
500 × 500	Steel mesh	13	27
750 × 750	Steel mesh	13	19

From Beaumont *et al.* (2003).

The resistance of an electrode is not changed by the applied waveform but does vary with differing values of ambient water conductivity. However, there is a fixed relationship between equivalent resistance values measured in different water conductivities (Equations (4.12) and (4.13)), so once it has been measured for a known conductivity, it can be recalculated for another.

$$R_{eq2} \div R_{eq1} = c_1 \div c_2 \qquad (4.12)$$

Therefore:

$$R_{eq2} = \left(c_1 \div c_2 \right) \times R_{eq1} \qquad (4.13)$$

where R_{eq2} is the resistance of the electrode in water of conductivity c_2; and R_{eq1} is the resistance of the electrode in water of conductivity c_1.

So, for example, if an electrode's equivalent resistance is 40 Ω in 350 $\mu S.cm^{-1}$ water, and we wish to determine its resistance in 800 $\mu S.cm^{-1}$ water, then from Equation (4.13):

$$R_{eq2} = \left(350 \div 800 \right) \times 40 = 17.5\ \Omega$$

An Excel spreadsheet (ElectroCalc) is available from the author that includes these calculations in the determination of power demand and projected voltage gradient from electric fishing set-ups.

4.6.2 Kirchoff's Law

The ratio of the resistance between the anode and cathode has a profound effect upon the voltage projected from the anode (Va). Kirchoff's Law states that the voltage across two resistors in a series circuit is proportional to the ratio of the resistance between them. Thus, the voltage actually propagated at the anode (Va) in an electric fishing set-up is not the same as the total circuit voltage (Vt), as is commonly supposed, but is proportional to the ratio of the anode-to-cathode resistance ratio. For example, if an anode and cathode both have equivalent resistances of 30 Ω and the total circuit voltage is 200 V, then the anode voltage will be 100 V (i.e. the Va will equal half Vt). However, if large anodes (e.g. 600 mm diameter rings) and small cathodes (e.g. a 750 mm long × 25 mm wide copper braid cathode) are used, this can result in anode voltage being as little as 30% of the circuit voltage (Beaumont *et al.* 2005).

Comparing the voltage gradient from a standard 600 mm anode with a range of cathode sizes shows the effect of optimising the resistance ratio between anode and cathode. Figure 4.30 clearly shows that for a fixed circuit voltage (Vt), the size of the projected anode field increased as a result of the increasing voltage at the anode (Va).

The effect is even more marked when using a twin-anode set-up. Many electric fishing control boxes have the capability to power two anodes; however, most only have one cathode socket. If an additional anode is connected and the

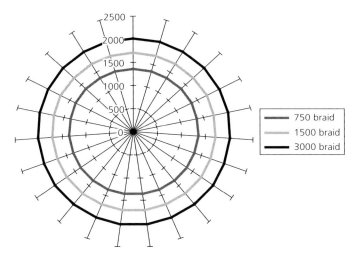

Figure 4.30 The effect of different cathode sizes on the distance to the 0.1 V.cm⁻¹ gradient from a 600 × 10 mm anode (axis measurements in millimetres).

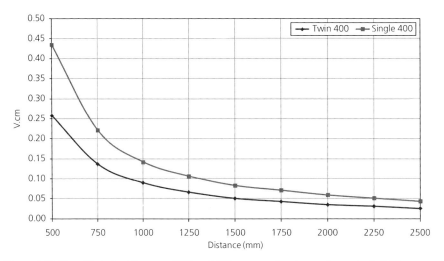

Figure 4.31 The effect of adding an additional 400 mm anode when using a single 750 mm braid cathode. With a constant circuit voltage (Vt) of 200 V, anode volts (Va) drop from 72 V to 43 V (from Beaumont *et al.* 2005).

cathode size not commensurately increased, a marked reduction in the voltage propagated from the anode can occur, enough in some circumstances to negate the effect of adding the additional anode (Figure 4.31). This effect has often been overlooked in instructions for setting up gear for electric fishing (a notable exception being Kolz *et al.* 1998).

It should be noted that as the voltage available to the anode increases, with the effect of increasing the capture radius, so too does the tetanising radius increase. Care needs to be taken, therefore, to either capture fish before they come too close to the anode or turn down the circuit voltage. The latter will also give overall power savings.

Kirchoff's Law can also be used to increase the anode voltage when, for example, fishing very low conductivity streams by increasing cathode length.

4.7 Conductance and conductivity

When dealing with the electrical resistance of water, it is more usual to express it as **conductance** (i.e. the ease with which an electrical current will pass through the water). Conductance is simply the inverse of the resistance $(1/\Omega)$, and it is measured in Siemens (S), although for many years it was measured in a unit called a 'mho', which is 'ohm' spelt backwards!

Conductivity is the ability of a unit volume of matter (e.g. a 1 cm cube of water) to conduct electricity. It is measured in Siemens per metre ($S.m^{-1}$), but in water it is usually measured in micro-Siemens per centimetre ($\mu S.cm^{-1}$). The conductivity of streams and rivers will vary depending upon the chemical make-up of the dissolved chemical ions (e.g. calcium) in it. The fewer the dissolved ions, the lower the conductivity, and the harder it is for electricity to pass through the water (distilled water is, in fact, an insulator). As conductivity values vary with temperature, roughly 2% per degree Celsius, values are normally corrected to the value they would be at 25 °C: this is called the **specific conductivity** (K). In electric fishing, however, we are interested in the conductivity at the ambient temperature and thus should use the uncorrected values or **ambient conductivity** ($K_{(t)}$). Formulae exist to correct K to $K_{(t)}$ (e.g. Mackereth *et al.* 1978).

$$K_{(t)} = K / \left(1.023^{(ts-ta)}\right) \qquad (4.14)$$

where *ts* = specific temperature (25 °C); and *ta* = ambient temperature. As an example, if the specific conductivity value was 800 $\mu S.cm^{-1}$, but the temperature was 6 °C, then the ambient conductivity would be 520 $\mu S.cm^{-1}$.

Generally, electric fishing can be carried out in water ranging in specific conductivity from 10 $\mu S.cm^{-1}$ to 5000 $\mu S.cm^{-1}$; by comparison, seawater is around 50,000 $\mu S.cm^{-1}$. The conductivity of the water that is fished has a big impact upon the electrical power required to energise the electrodes and the effect the voltage gradient (power gradient) has on the fish (discussed further in this chapter). Alabaster and Hartley (1962) showed a clear, positive relationship between water conductivity and sampling efficiency. They considered the relationship was due to the expansion of the effective capture area of the electric field (at a fixed circuit voltage) of the electrodes at the higher conductivities (i.e. the voltage gradient required to immobilise the fish was lower at the higher conductivity). Likewise, Pusey *et al.* (1998) considered that differing first-pass fishing efficiency between two rivers were due to conductivity differences. However,

Penczak *et al.* (1997) found that electric fishing efficiency indices were not significantly correlated with increasing conductivity. Nevertheless, Penczak *et al.* (1997) found that the capture rate and the catch efficiency index were highest at the highest conductivity.

4.7.1 High-conductivity water

Very high-conductivity water (above 5000 µS.cm^{-1}) can be sampled using electric fishing, but the power and current demand are very large and require extremely high-output generators. The use of short-duty-cycle pDC waveforms will reduce the power demand, and increasingly marine trawls have electric 'tickler' arrays to lift fish from the seabed. These electrical devices markedly increase capture rates of bottom-living fish and 'shrimps', reduce the number of 'discards' in the catch, reduce damage to the sea floor and (perhaps most importantly for the operators) reduce fuel costs of the boats by as much as 50% (Soetaert *et al.* 2015). Research in the 1950s indicated that marine fish could be drawn to electrodes with an exponential waveform of 20 Hz pDC using a peak voltage of 2000 V (10,000–15,000 A). Bary (1956) found that in seawater, DC waveforms required four times more power than AC to induce electronarcosis. AC waveforms were, however, more severe in their effect on the fish, with substantially longer recovery times needed. Bary (1956) also found that in seawater (33$^0/_{00}$ salinity), exponential pDC gradients of 0.4–1.1 V.cm^{-1} at frequencies of 15–30 Hz and pulse duration of 4–115 µs were effective in inducing electronarcosis in mullet (*Mugil* spp.). Nordgreen *et al.* (2008) found that a 1 s exposure to 50 Hz DC at 0.033 V.cm^{-1} 'stunned' all of the test Atlantic herring in seawater; however, a high proportion of the fish had fractured spines. Whilst little research on freshwater-style fishing in seawater has been carried out, some experimental work has recently been published on prototype equipment that will work in very high-conductivity (20,000 µS.cm^{-1}) water (Warry *et al.* 2013). Exact details of the electrical set-up are not given, but the equipment probably uses very low voltage (Warry *et al.* (2013) stated less than 200 V, but for the high conductivities quoted, voltages lower than that could be used – see Section 4.5.3). The equipment probably also used an exponential pulse waveform to limit the mean power demand; even so, amperages of greater than 50 A are quoted. Equipment of this type would enable 'transitional' waters, as defined by the EU Water Framework Directive, to be fished and thus expand considerably the habitats where fish population structure can be determined.

4.7.2 Low-conductivity water

Very low-conductivity water also creates problems for electrofishing (e.g. Alabaster & Hartley 1962, Allard *et al.* 2014). Distilled water (conductivity 0 µS. cm^{-1}) is an insulator: with no free ions, electricity cannot be conducted through the medium. Sternin *et al.* (1976) reports that 5 µS.cm^{-1} is the lowest water conductivity in which it is possible to electric fish. In low-conductivity water (less than 50 µS.cm^{-1}) where there is little ionic content to the water, very high-voltage gradients are needed to provide sufficient energy (per charged particle) to propagate the electric field. These problems are in addition to effects of the fish–water conductivity effects (see Section 4.8.1). At very low conductivities (e.g. below 30 µS.cm^{-1}), DC voltages in excess of 1000 V may be needed, and often pDC waveforms are needed in order to utilise the lower attraction and immobilisation voltage thresholds of this waveform. As the curve of the relationship between water conductivity and resistance is very steep below 30 µS.cm^{-1}, even small changes in conductivity have a big impact on the voltage gradients required to immobilise fish. In water below 20 µS.cm^{-1}, pDC voltages of 990 V have been found to be required to immobilise the fish sufficiently for capture. Even then, the population capture field was small, and population assessment (using depletion methods) was poor (compared with results from rotenone poisoning); however, the measure of species diversity can be good (personal experience of the author). Adjusting anode diameter to produce a smaller, more intense field and increasing cathode length to maximise anode voltage (from Kirchoff's Law) can also help maximise the likelihood of fish capture under these marginal conditions.

4.8 Fish conductivity

As well as the problems of electric field propagation and power demand in varying water conductivities, the conductivity of the water has been found to affect the electrical output needed to incapacitate the fish at differing water conductivities. The underlying cause for the variation is the concentrating or dissipating effect of the electric field depending upon the ratio between the conductivity of the fish and water.

Unfortunately, when it comes to determining fish conductivity, differing and nonstandardised methods have been used to determine the conductivity of fish. The data that exist suffer from variation in technique used to measure the values, making comparisons between different studies difficult. Several historic studies have been made on the conductivity of fish using electric circuit theory and fish immersed in known-conductivity water, so-called 'true' or 'classical' conductivity. Halsband (1967) found values of between 800 to 1200 µS.cm^{-1} for four different fish species (Table 4.4). Tests by the author, using 'classical' conductivity methods, gave the conductivity of an Atlantic salmon smolt of about 1500 µS.cm^{-1}.

Resistivity fish counters use electric circuit theory to determine when fish are present across the electrodes. The counting electrodes are energised with a low voltage (around 5 V) and the electrical current through the circuit monitored. When fish are present across the electrodes, the bulk resistance of the water over them alters, causing an increase in electrical current. On the river Frome in

Table 4.4 Fish ('true') conductivity.

Species	Conductivity (μS.cm^{-1})
Salmon (*Salmo salar*)	1550
Trout (*Salmo trutta*)	1220
Perch (*Perca fluviatilus*)	1089
Carp (*Cyprinus carpio*)	870
Gudgeon (*Gobio gobio*)	814

Derived from Halsband (1967) and author's data.

Table 4.5 Variation in ambient fish ('true') conductivity with temperature.

Temperature (°C)	Ambient conductivity (μS.cm^{-1})
5	372
10	543
15	714
20	1026
25	1969

Whitney and Pierce (1957).

Dorset, UK (mean ambient water conductivity around 450 μS.cm^{-1}), the presence of a fish over the electrodes has the effect of reducing the resistance of the water (i.e. the fish is more conductive than the water). This resistance reduction is clearly seen by the fact that the current increases, not decreases. At other sites with higher water conductivity, the fish may increase the resistance and thus current will decrease, depending on what the conductivity of the fish and water are. Smolian (1944) also considered that fish always conducted electricity better than the surrounding water (i.e. had a higher resistivity).

It is possible that the variable capture efficiencies or damage susceptibility of electric fishing on different species, often observed by operators, could be due to the variation in fish conductivity between different fish species.

Fish conductivity (however measured) will vary with temperature. Whitney and Pierce (1957) determined conductivity of a carp varied between 400 and 2000 μS.cm^{-1} at temperatures between 5 and 25 °C (Table 4.5).

If the values in Table 4.5 are correct, it has important implications regarding the settings used for electric fishing at different seasons. For example, lower voltages may be needed to minimise fish damage in the summer.

Sternin *et al.* (1976) found that, when measured with a pulsed current, the apparent conductivity of the fish increased with increasing frequency of the pulses; they attributed this to the decrease in capacitance of the cell membrane of the fish skin. They found differing conductivity values for different species of fish and stated that 'differing species fish conductivities require different power values to create an effect'. Sternin *et al.* (1976) also considered that the scale type and skin of the fish also affect the conductivity, as a result of the low conductivity of the scales 'shielding' larger scaled or mucous-skinned fish from the electrical current.

Kolz and Reynolds (1989), in a radical departure from circuit theory, used the behavioural response reaction of fish to an electric field to determine what they described as the conductivity of the fish. When fish reaction to a voltage gradient was at its maximum (i.e. the point where maximum behavioural response was achieved with the minimum power transferred into the fish), they considered that the conductivity of the fish and water were the same. The conductivity values obtained by this method are substantially lower than those determined from circuit theory; Kolz and Reynolds (1989), for example, quoted values of effective conductivity for goldfish of 69–160 μS.cm^{-1} with an average of 125 μS.cm^{-1}. Recognising this discrepancy, Kolz and Reynolds (1989) renamed this measure of conductivity the 'Effective Conductivity' of the fish. Further exploring the differences between conductivities, Kolz (2006), using measurement of electrical current in conductive rubber cylinders, defined the conductivity of an object submerged in water as 'Immersion Conductivity'. In this work, power density actually transferred into the cylinder was measured using probes inserted into cylinders. The work also established both the relationship and the reason for the discrepancy between 'Classical' and 'Immersion/Effective' conductivity measurements. Preliminary experiments on two dead goldfish showed similar values between (the empirical) immersion conductivity and (the behavioural) effective conductivity, with mean fish conductivity ranging between 100 and 150 μS.cm^{-1}. The work also observed the importance of boundary conditions around the object (i.e. the importance of the cross-sectional area of surrounding water compared with the cross-section of the object). Kolz (2006) stated that the variation is expected, although how an operator is meant to set up specific electric fishing output with such variable parameters is not made clear.

Further studies (Burkardt & Gutreuter 1995, Miranda & Dolan 2003) indicate that the immersion fish conductivity values proposed by Kolz and Reynolds (1989) may not vary greatly between different species.

4.8.1 Water–fish conductivity ratio

Common to all methods of assessing fish conductivity is that applied voltages need to be varied to attain good capture rates and minimal injury at differing conductivities of water. In broad terms, at low conductivities the electrical output applied into the water needs to be increased, and in high-conductivity water the voltage can be decreased. Cuinat (1967) considered that between conductivities of 500 μS.cm^{-1} and 50 μS.cm^{-1}, applied voltages should increase by 100 V each time the conductivity halved (for DC with an anode potential of 300 V at 500 μS.cm^{-1} as a starting point). Lamarque (1967) also considered that voltage gradient should be increased for DC in differing (lower) water conductivity. In situations where small streams are being fished, using a smaller anode can (for the same circuit voltage) increase the voltage gradient in close proximity to the electrode (Allard *et al.* 2014), thus improving its fish catching properties at that location. These adjustments bring with them problems of their own with regard to the safety of using high-voltage electricity (>900 V may be required). Under extremely low conductivities (<20 µS.cm⁻¹), salt can be added to the stream to increase water conductivity (Lamarque 1990) and thus reduce the voltage gradients needed to incapacitate the fish. At high conductivities, it is possible to reduce the voltage gradient and still get satisfactory fish capture efficiency. Broad guidelines for settings that have been found to be effective at different conductivities are given in Table 4.6: based on a 300–400 mm anode and 3000 mm cathode and powered by an AC generator of maximum 3.5 kVA output with an appropriate power factor.

Kolz and Reynolds (1989) provided information on the voltage gradients required to create narcosis for both DC and pDC at different conductivities (Figure 4.32). These are independent of electrode configuration, so the output of

Table 4.6 Suggested circuit voltages for differing-conductivity water (based on a 300–400 mm anode and 3000 mm cathode and powered by an AC generator of maximum 3.5 kVA output). Note: frequency and duty cycle used will also affect circuit voltage needed.

Water conductivity (µS.cm⁻¹)	Applied voltage (V) pulsed DC	Applied voltage (V) DC
Less than 50	>500	>700
50–150	300–400	400–500
150–500	150–250	300–400
500–800	150–200	Too high power demand
800–1000	120–180	Too high power demand
More than 1000	50–150	Too high power demand

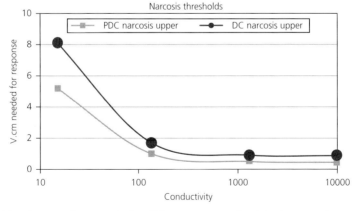

Figure 4.32 Voltage gradient required in differing water conductivity for DC and pDC waveforms (derived from Kolz and Reynolds 1989).

any anode–cathode combination can be measured, and the correct gradient 'dialled in'.

There are three different ways that have been used to describe the reason for requiring the electrical output to be varied, and each will be discussed in turn:
Graphic depiction
Circuit theory
PTT.

4.8.1.1 Graphic depiction

Figure 4.33 shows a graphic depiction (as described by Brøther 1954) of the principles behind the different voltages experienced by a fish at differing water conductivities. The horizontal lines represent lines of electrical current, whereas the two lines from either end of the 'fish' are the voltage equipotentials (which are always at right angles to the electrical current).

In Figure 4.33a, the conductivity of both the fish and water are the same, and the voltage experienced by the fish will equal that measured in the water if the fish were not present (e.g. 5 V).

In Figure 4.33b, the conductivity of the fish is greater than that of the water, making it easier for the current to flow through the fish rather than round it. This distorts the electrical field, and the fish experiences a lower head-to-tail voltage gradient.

In Figure 4.33c, the fish has a lower conductivity than the water, it is easier for the current to flow round the fish and therefore the fish receives a higher head-to-tail gradient.

4.8.1.2 Circuit theory

The presence of a fish in a body of water can be envisaged as two resistors in series in an electrical circuit, with R_w being the resistance of the water and R_f the resistance of the fish. The total circuit resistance (R_T) therefore is $R_w + R_f = R_T$ (Figure 4.34).

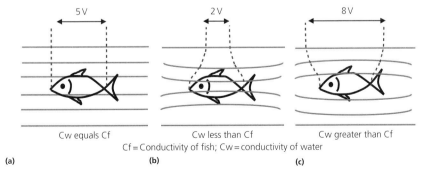

5 V	2 V	8 V
Cw equals Cf	Cw less than Cf	Cw greater than Cf

Cf = Conductivity of fish; Cw = conductivity of water

(a)   (b)   (c)

Figure 4.33 Distribution of electrical current and voltage in similar and dissimilar conductive media (derived from Bröther 1954).

Figure 4.34 Schematic diagram of fish and water forming a simple series resistance circuit.

According to Kirchoff's Law (see Section 4.6.2), the voltage across each resistor is proportional to the ratio of their resistance. If the circuit voltage, the resistance of the water and the resistance of the fish are known, then Ohm's Law can be used to determine the voltage 'across' the fish.

For example: if the water conductivity is 200 μS.cm^{-1} (5000 Ω) and the fish conductivity is, for example, 125 μS.cm^{-1} (8000 Ω), then circuit current can be determined by:

$$I = V / R_T = 100/(5000+8000)=0.008\,A$$
$$\therefore Vw = Rw \times I = 5000 \times 0.008 = 40\,V$$
$$\&Vf = Rf \times I = 8000 \times 0.008 = 64\,V.$$

where I = circuit current; R_T = the total circuit resistance; Vw = the voltage in the water; and Vf = the voltage across the fish.

However, if the water conductivity is 2000 μScm^{-1} (500 Ω) and the fish is 125 μScm^{-1} (8000 Ω), then:

$$I = V / R_T = 100/(500+8000)=0.012\,A$$
$$\therefore Vw = Rw \times I = 500 \times 0.012 = 6\,V$$
$$\& Vf = Rf \times I = 8000 \times 0.012 = 96\,V.$$

As can be seen, voltage experienced by the fish, and thus the reaction of the fish, for the same circuit voltage will vary depending on the conductivity of the water.

4.8.1.3 Power Transfer Theory (PTT)

Kolz and Reynolds (1989) observed conflicting responses between voltage gradient (E, volts.cm) and current density (J, amps.cm^2) needed to produce a 'stun' response in goldfish. They considered that it is the magnitude of the power or power density (D, watts.cm^3), a parameter that integrates E and J (Equations 4.15 and 4.16), which determines the success or failure of electric fishing. To explain the principle, they proposed the concept of PTT where specific fish reactions (twitch, stun etc.) only occur after a specific, fixed amount of power has been transferred into the fish:

$$D = C_w \times E^2 \tag{4.15}$$

or:

$$D = J^2 \div C_w \tag{4.16}$$

where D is power density; E is voltage gradient; J is current density; and C_w is water conductivity.

A diagrammatic representation of the concepts of voltage gradient (E), current density (J) and power density (D) is shown in Figure 4.35. Whilst E can be measured directly, J and D need to be calculated.

PTT proposes that the effect of an (in-water) electric field on fish is maximised when the fish and water are the same conductivity, that is, the mismatch ratio (of the water and fish conductivity) equals 1. When the conductivities differ (water conductivity is either greater or lesser than that of the fish, and the mismatch ratio ≠ 1), then a mismatch occurs, power transfer and effect on the fish are reduced and applied power density will need to be increased over that required where the mismatch ratio is 1 (Figure 4.36).

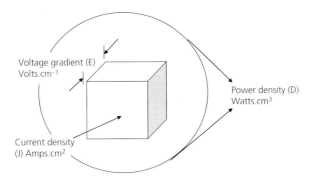

Figure 4.35 Diagrammatic representations of the three electrical values used to describe the properties of the power of electric fishing fields.

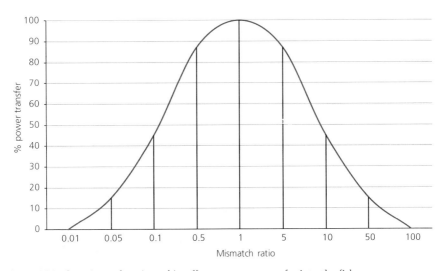

Figure 4.36 The mismatch ratio and its effect on power transfer into the fish.

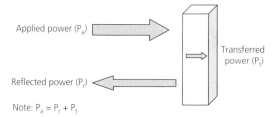

Applied power (P_a)

Transferred power (P_t)

Reflected power (P_r)

Note: $P_a = P_r + P_t$

Figure 4.37 A schematic representation of the principles of Power Transfer Theory.

Kolz and Reynolds (1989) used wave theory to explain the reason for the process. Using a model of electrical power flow at the boundary of two resistive media, they broke the power into three components: applied power (P_a), transferred power (P_t) and reflected power (P_r) (Figure 4.37). The results show again that maximum power is transferred (to the fish) when resistances are equal.

The advantage of PTT is that it gives a nonsubjective method whereby electric fishing output can be standardised for differing ambient water conductivity to give directly comparable capture efficiency data (Berkhardt & Gutreuter 1995). For agencies that require categorising the status of their fish stocks, this is an important breakthrough as population density results need to be comparable between sites irrespective of parameters apart from the actual fish numbers.

According to Kolz (1993), in water below the 'immersion conductivity' of the fish (according to his calculations, ~150 µS.cm⁻¹), voltage should be increased. Above the 'immersion conductivity' of the fish, current (for which he suggests pulse width) can be increased. This results in power transfer to the fish remaining constant.

In order to calculate appropriate output settings for differing-conductivity waters, PTT does require knowledge of output settings (either output voltage and output current or voltage gradients) that, for a defined conductivity of water, produce good capture rates with minimal injury. Unless voltage gradient parameters are used, these settings will be appropriate only for the gear (i.e. anode and cathode size) that determined the optimal settings. The following shows the process of calculating output at different conductivities. These examples are based on the calculations in Kolz *et al.* (1998) and have used a 400 mm anode, a 3000 mm cathode and ambient water conductivity (C_w) of 350 µS.cm⁻¹ giving an electrode resistance value (at 350 µS.cm⁻¹) of 46 Ω. This set-up when operated using pDC at 200 V, 50 Hz, 25% duty cycle results in a peak power output of 870 watts. The set-up has been shown to give good capture and very low injury rates for juvenile Atlantic salmon parr (author's experience). Effective fish conductivity (C_f) is taken as 150 µS.cm⁻¹, based on Kolz and Reynolds (1989).

Firstly, determine optimal settings – as above – by experience or trial and error.

1 Calculate the mismatch ratio (q) between the conductivity of the fish and water:

$$q = C_w / C_f = 350/150 = 2.33$$

2 Calculate the power conversion factor (PCF). This is the function that accounts for the power actually transferred into the fish in Kolz and Reynolds (1989); it is called the multiplier for constant power (MCP).

$$PCF = \left[(1+q)^2 /(4q)\right] = (1+2.33)^2 /(4 \times 2.33) = 11.09/9.32 = 1.19$$

3 Calculate the power goal (P_g) at matched conditions (i.e. the transferred power into the fish at the C_w at which the good capture rate was observed).

$$P_{g(Cw=350)} = \left[P_{(Cw=350)} /PCF\right] = 870/1.19 = 730 \, Watts$$

Now that the power goal is known, settings to achieve this 'standardised' power goal can be calculated for any conductivity water. For example, if water of $100 \, \mu Scm^{-1}$ is fished:

1 Calculate the mismatch ratio:

$$q = C_w /C_f = 100/150 = 0.67$$

2 Calculate the PCF:

$$PCF = \left[(1+q)^2 /(4q)\right] = (1+0.67)^2 /(4 \times 0.67) = 2.79/2.68 = 1.04$$

3 Calculate the power goal (P_g) required at $100 \, \mu Scm^{-1}$. This time, however, the matched power goal needs to be *multiplied* by the PCF:

$$P_{g(Cw=350)} = \left[P_{(Cw=350)} \times PCF\right] = 730 \times 1.04 = 759 \, watts$$

In order to calculate the new voltage settings, the electrode resistance needs to be calculated for the new conductivity:

$$R_{100} = \left[(C_{w350} /C_{w100}) \times R_{350}\right] = \left[(350/100) \times 46\right] = 161 \, \Omega$$

From this, the appropriate voltage output can be calculated:

$$V = \sqrt{(P \, X \, R)} = \sqrt{(759 \times 161)} = 350 \, Volts$$

Changing the output voltage from 200 to ~350 V should therefore give equivalent capture probability (to the $350 \, \mu S.cm^{-1}$ water) in the $100 \, \mu S.cm^{-1}$ water.

It should be noted, however, that the theory has not received unanimous acceptance by fishery researchers. Snyder (2003) debated whether power can actually be 'transferred' into a fish (whereas current certainly can). Henry and Grizzle (2006) found that mortality (caused by exposure to electricity) could be predicted more accurately by variability in peak E and mean J than with models that use the effective (immersion) conductivity of the fish. Problems also occur when considering whether the effect is working on peak or RMS values, particularly for pDC waveforms (Beaumont *et al.* 2000); if the effect is caused by peak values, it is not clear how increasing the duty cycle (whilst keeping the peak voltage constant) improves the effect. However, the data from Sternin *et al.* (1976) would tend to corroborate the principles of PTT, that threshold values are defined by the product of voltage gradient and current density, that is, power density. The great strength of PTT is that it gives a standardized method of compensating for different capture rates at different water conductivity, or just simply determining an output voltage that will actually catch fish in low-conductivity waters. Even if the theory still has uncertainties, this must be considered beneficial for researchers and managers.

CHAPTER 5

Electric fishing equipment

Regulations regarding the construction of, and safety features on, electric fishing equipment range from stringent to none. In the countries covered by EU regulations, all equipment built specifically for electric fishing should comply with European Standard 'Safety of household and similar electrical appliances – Part 2: Particular requirements for electric fishing machines' (Anon. 2003b). Many individual countries within the European Union and organisations within those countries (e.g. the UK Environment Agency) then have further regulations regarding the construction and use (particularly relating to safety features) of that equipment.

During the author's time fishing in different countries, I have seen equipment that was so constrained by safety features that it was almost impossible to use it (e.g. tilt switches on backpacks constantly tripping with just normal walking movement, anode switch settings being so fine-tuned that the system constantly went into error status, and double-pole dead-man's relay switches having to be synchronised to millisecond precision). In contrast, I have seen equipment where there was a real risk of injury to operators from using the equipment – issues such as getting operators to put (an insulating) plastic fertilizer sack on their back before putting on the backpack gear (to avoid issues with the metal frame becoming live), or either no dead-man's switch or domestic (non-water-proof) doorbell push switches being taped onto (wooden) anode poles as dead-man's switches. There can also be issues with organisations trying to specify build criteria instead of safety criteria. One example is specifying that batteries should be in waterproof housings, when what is actually required is that the battery and its connection to the pulse box are waterproof; this can be done by methods other than by putting them in a waterproof box.

Whilst it is important that equipment is intrinsically safe to use, it is also important that it is actually usable. I would never wish to go back to the equipment I used when I began my career (no safety features at all). However, good training of operators and 'Safe Working Practices' should minimize risk and still allow sensibly designed equipment to be usable.

Electricity in Fish Research and Management: Theory and Practice, Second Edition. W.R.C. Beaumont.
© 2016 John Wiley & Sons, Ltd. Published 2016 by John Wiley & Sons, Ltd.

Irrespective of safety, even experienced operators will struggle if using poorly ergonomically designed equipment. For example, the ergonomics of the anode design can make a difference to the ease of use and thus increase the likelihood of fish capture. Some manufacturers have produced pulse boxes where a cable gland is positioned next to the equipment handle, resulting in difficulty carrying the box; others have cathode exits on the opposite face to the handle, so the box is put down on this socket. Heavy equipment can also cause fatigue and possible injury when lifting and transporting, and due regard should be given to local Health & Safety guidance on maximum limits for safe lifting.

It is important that all uses of electric fishing equipment are familiar with, and comply with, the regulations in force in their country, state or organization.

5.1 Generators

Generators used for electric fishing can be either separate 'stand-alone' units into which a control box is plugged or units where the generator and control box form a single integral unit.

In the latter case, the energising voltage is often taken directly from the generator windings to provide different, fixed, circuit voltages.

Historically, smooth DC, single-phase AC and three-phase AC generators have all been used to power electric fishing equipment (Hartley 1980b). With the development of modern lightweight single-phase AC generators (particularly those using inverter technologies), most are now single-phase 230 V AC units. They should be modified to have isolated outputs (i.e. the 'neutral' output is not referenced to 'Earth' or the framework of the generator) (Hartley 1975, Goodchild 1990). Because of this, commercial or industrial generators modified for use in electric fishing must not be used for any other purposes and should be clearly labelled 'For Electric Fishing Only'.

Generator capacity can be specified in volt-amps (VA) or kilovolt-amps (kVA), although some manufacturers specify watts (W) or kilowatts (kW). The two measurement units are related by the *power factor* (see Section 4.1), and only in the special case of a *unity power factor* (i.e. where the power factor is 1.0) are they equivalent. It is important that the rating of the generator is not exceeded to ensure reliable operation, to avoid damage to the generator and to avoid break-down of the smooth AC waveform that should be being produced. If smooth AC is not delivered to the pulse box from the generator, then this can result in very uneven or spiky pDC waveforms being produced from the pulse box (Beaumont *et al.* 2002).

Normally, it is simply necessary to ensure that the generator capacity is adequate to undertake the fishing operation under consideration. In electrical terms, there is no disadvantage in using a generator that is larger in capacity than required because the actual power supplied by the generator in any situation is determined purely by the output voltage of the generator (constant) and by the applied load. In practical terms, however, it is normally desirable to use the smallest adequate generator, therefore minimizing the size and weight of the equipment to be carried to the fishing location.

The power required for any particular electric fishing operation will depend upon the size, number and position of the electrodes, the voltage applied to the electrodes, the waveform used and the conductivity of the water. In general, the higher the conductivity, the greater the generator power required. Water depth and width do not influence power requirements (assuming the number and size of the anodes remain the same). Conductive riverbeds will affect the power loading by decreasing the equivalent resistance of the cathode (assuming it is lying on the riverbed).

In summary, the minimum capacity of the generator for any electric fishing operation will depend upon three factors:

1 The power dissipated in the water
2 The conversion efficiency of the electric fishing control box
3 The power factor of the control box.

The **power** in the water can be determined by the appropriate formulae (Section 4.5), with the water representing a purely resistive, linear load and the combined electrodes and water forming the equivalent resistance of the electrodes.

The **conversion efficiency** of the electric fishing control box will depend upon the design and circuit of the power converter. Power losses may, for example, be due to switching losses, conduction losses, transformer losses, control circuit losses and so on. A well-designed modern unit should be expected to have a conversion efficiency of up to 90%.

An adverse **power factor** occurs in input power converters of the type used in AC generator–powered electric fishing control boxes, due to the nature of the circuits used when AC is converted to DC or pDC. Due to power losses within these circuits, the control box actually uses more power than simple calculations would suggest. See Section 4.5.1 for more information on the power factor.

5.1.1 Use of multiple generators and control boxes

Under certain circumstances (very wide water systems), it may not be possible to have sufficient anodes powered from one generator or width of field from a single boom-boat to cover the water. Under these circumstances, multiple fishing arrays powered by multiple power sources with associated multiple control boxes have been operated. Extremely vigilant safety systems should be put in place when using such set-ups, as there is a high possibility of someone who falls or slips in the water being electrocuted by equipment that is being used some distance away from them by operators who are unaware of their predicament.

Although more research is required, there is also the possibility of problems with waveforms from control boxes powered from different generators interacting

Figure 5.1 Multiple generator and control box fishing. (With permission of N. Poulet, ONEMA France.)

and producing higher (or lower) than expected voltage gradients, or out-of-phase pulsed waveforms creating high-frequency (damaging) waveforms.

The UK Environment Agency regulations prohibit the use of more than one power source in the water at any time. Figure 5.1 shows multiple units being used to intensively fish a stream. W.G. Hartley's comments (Hartley 1975) about the risk increasing with the number of operators seem particularly appropriate.

5.2 Control boxes

The source electricity used in electric fishing is conditioned and adjusted by a control or pulse box before being used at the electrodes. All control boxes should be fitted with emergency off buttons that instantly shut off power to the electrodes.

Control boxes will be rated according to the maximum current loading (amps) at which they can operate. In the past, this loading was always the root mean square (RMS) or average loading, but recently some boxes have been using peak loading as the limiting factor. These peak loads are often at relatively low values (e.g. the 'Easyfisher' box (EAES Ltd. UK) uses 8 A peak loading, whereas an Electracatch FB4A box (Electracatch UK) is rated at 4 A mean). For a square-wave pDC waveform at 25% duty cycle, this would equate to 16 A peak. The low-value peak amp limited boxes are not suitable, therefore, for fishing in high-conductivity waters, particularly if using twin anode systems (note: EAES Ltd have recently uprated the amp limit on their boxes). For operators wishing to use DC outputs in moderate to high-conductivity water, control boxes need to have extremely high current ratings. For example, if fishing 500 μS.cm^{-1} water using DC and a single 400 mm anode with a 3000 mm cathode, then the current demand will be around 13 A, necessitating at least a 20 A control box to allow a safety margin. In 1000 μS.cm^{-1} water, a 60 A rated control box would be needed.

For a given voltage, high-conductivity water will have a higher current loading than lower conductivity water (amps being a function of voltage and resistance). It is important to know the loading that will be put on the boxes at the conductivity being fished in order not to overload the gear with possible dangerous effect.

Older, and where local codes of practice allow, control boxes may have the capability of switching between some or all of AC, DC and pDC output waveforms. Waveform pattern output from the control box should be known by operators so that they can understand potential fish welfare problems and understand what variations in output parameters are available. For example, operators should know the pulse type being used when on pDC setting (square, half-wave rectified etc.). They should also know how the pulse shape reacts to changes in water conductivity and at different settings. Under differing power conditions, either by design (to limit power to components) or due to the limitations of the components themselves, pulse shape may change when being fished at high loadings. For example, in the older designs of one UK manufacturer's control box, the waveform changes from full, half-wave rectified pulses to quarter-wave pulses depending on the power output. In some boxes, the voltage is automatically reduced when the current limit is reached; this protects the box but will alter the catch characteristics of the unit.

Ideally, all pulse boxes should have clear information and oscilloscope waveforms showing how they operate under varying conditions of load so that the effects of adjustments in output can be clearly understood.

A wide variety of controls and meters have been fitted to control boxes over the years, and it is important to know what the controls do and what the meters actually display. In particular, as voltmeters may either show input voltage from the generator or voltage (RMS or peak) being output to the electrodes or have dual function, it is important that operators know and understand what is being displayed.

Safety circuits also vary between manufacturers and age of equipment, and on some older boxes there are no emergency cut-out buttons. In addition, on some boxes, touching the anode and cathode together can catastrophically affect the internal electronics. Manufacturers' manuals should be carefully read and limitations of the gear understood before equipment is used.

5.2.1 Generator-based control boxes

Most generator-based control boxes are designed to be located in a fixed position on the river bank or within a boat and energise the anodes via electrical cables from the river bank or boat. There are a few units, however, that are designed to be worn by the operator (Figure 5.2). Whilst these units have a useful role if fishing moderate to high-conductivity water where there is no easy access for transporting a bank-based set-up, they are noisy and create noxious exhaust fumes near the operator. In addition, local safety regulations may prohibit wearing a running generator due to safety concerns about the hot exhaust and fuel tank.

Figure 5.2 A backpack generator and control box fishing unit.

Control boxes that are designed to be used with generators can be divided into three broad categories:
1 Ones with no facility to control output
2 Ones with limited ability to control output
3 Ones where many parameters of the output can be controlled.
This sequence of sophistication is related to the increasing technological development of the electronics used in the boxes. This has been driven by the increasing awareness by operators of the importance of using fish-friendly output.

5.2.1.1 Control boxes with no facility to control output
Boxes in this category (Figure 5.3) may work well and give good capture and fish welfare rates for particular sites. However, use of the boxes in different and differing conductivity waters will result in unequal catch probability between sites (important if single-pass surveys are being undertaken) or damaging outputs.

5.2.1.2 Control boxes with limited ability to control output
This type of box (Figure 5.4) typically has a single 'Power' control. An ammeter to show circuit current may also be fitted (see Section 4.4 for the effect that differing water conductivity has on amps used). On some boxes, a voltmeter

Figure 5.3 Control box with no output controls.

Figure 5.4 Example of a pulse box with limited control of output, and a close-up of the controls and instrumentation.

may be fitted; however, it often only shows the input voltage from the generator, not the output volts to the circuit.

The 'Power' control often alters both the peak voltage of the pulse and the pulse width. This is achieved by varying the lower voltage at which the full-wave (for 100 Hz pulses) or half-wave (for 50 Hz pulses) is truncated. Increasing the power therefore increases both the peak volts and the pulse width. In an ideal situation, these two parameters should not be linked in electric fishing, as in high-conductivity water lower peak voltage but higher pulse width is required, and in low-conductivity water high peak voltage but lower pulse width is required. Figure 5.5 shows the effect of increasing the power on one of these types of control box. The voltage (pulse height) increases, but so does the pulse width.

On other boxes, the waveform may change (from quarter-sine to half-wave rectified) as the power is increased. This is not ideal due to concerns regarding the differing severity that the two waveforms have in their effect on fish. Despite

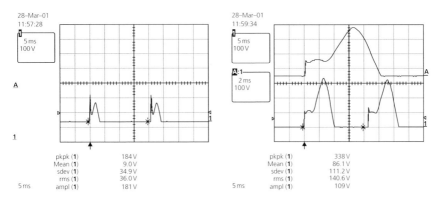

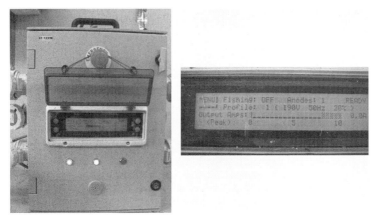

Figure 5.5 Effect of increasing 'power' control on a Millstream FB3A control box. Both tests are at 50 Hz and 200 V generator input with 52.6 Ω load. Note the uneven waveform caused by poor generator output waveform.

Figure 5.6 An example of a control box where all output parameters are adjustable via a liquid crystal display (LCD) and menu settings.

these problems, these boxes work well in medium-conductivity systems once a suitable output setting has been established – often by trial and error.

5.2.1.3 Control boxes where many parameters of the output can be controlled

Most modern control boxes come into this category, where operators can adjust output voltage, pulse width and pulse frequency independently of one another (Figure 5.6). The ability to control output enables very precise control of settings based on knowledge of site and equipment characteristics. It does, however, also require that operators are trained to understand the principles behind electric fishing and the parameters used.

Some manufacturers produce generator-based control boxes that enable very high voltages to be used (600 V). This is sometimes achieved by having a separate transformer that plugs into the control box. Such systems are very heavy and only really suitable for boat-based fishing in large low-conductivity waters.

5.2.2 Battery-powered control boxes

On these units, output is usually confined to DC or pDC. If the output is pDC, waveforms are usually either exponential discharge or square wave. Exponential waveforms are often used due to the low power demand of this waveform giving much longer battery longevity (Beaumont *et al.* 1999, 2000). On some equipment, there is the capability to switch between DC and pDC depending on how the anode safety switch is operated (e.g. Hans Grassl IG600 units); this optimises both battery longevity and fish welfare. Operators fish using pDC, and when fish are seen to be drawn within the electric field, the output is switched to DC to improve fish welfare near the anode. In actual field use, however, with multiple fish present, this is probably not a practical proposition. Some boxes have the capability to create gated bursts or other novel pulse patterns: the Smith-Root Inc. backpack box has the capability to produce 256 different outputs. However, many of the novel outputs have limited practical application in field conditions.

Whilst some bank-based battery systems exist, most are designed to be carried (worn) by the operator. Equipment can either be wholly fitted into a backpack type of carrier (Figure 5.7a) or split with the control box worn on the chest and the battery pack worn on the back (Figure 5.7b). Due to the potential dangers of wearing the control box, modern backpack equipment should have additional safety circuits (compared with generator-based units). These will often consist of immersion sensor and tilt switches that are designed to shut off power if the operator goes in too deep in the water or trips or falls into the water. Operators have mixed views of certain designs of tilt safety systems. Systems can be too

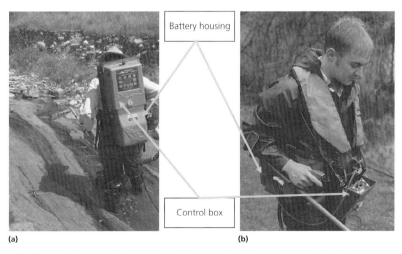

(a)   (b)

Figure 5.7 Examples of (a) back-mounted and (b) front-and-back-mounted battery-powered electric fishing units.

sensitive, shutting off the power just on the sudden movement of the operator and making the gear very inefficient: there is often a fine line between safe and unusable. Other equipment acknowledges the practicalities of using the equipment and allows a greater degree of front tilting compared with backwards tilting, allowing operators to stoop under obstruction without the unit tripping off.

Batteries should be of non-spill, sealed design and are usually gel-type lead–acid batteries. Because of the weight of the lead–acid batteries used, gear can be very heavy (15 kg). Using smaller, lower capacity batteries (with consequent lower fishing duration) will reduce this weight, but batteries will need to be changed more often. In recent years, batteries using lithium iron phosphate (instead of lead) have become available, and these can be considerably lighter (for the same output) than lead-based batteries. They are, however, considerably more expensive.

5.3 Electrodes

'Electrode' is the term applied to the bare metal contacts (both the positive anode and the negative cathode) through which the electricity is conducted into the water. When electric fishing, electrical power must cross two boundaries, one from the electrodes into the water and one from the water into the fish (Kolz 1993). The electrode design (shape, construction material etc.) and its consequent characteristics are important factors in achieving the propagation of electric fields into the water, thus creating the required physiological effects upon the fish.

The geometric configuration (shape) of the individual electrodes defines the shape, size and distribution of the voltage gradient and therefore the electrical current and power gradient in the water (Kolz 1993). This voltage and power distribution are further affected by electrical field boundary effects, that is, electrode placement in the water (surface or bottom of the water) or other boundary effects such as the bank of the river or lake or highly conductive sediment or bank reinforcing (e.g. steel piling). The electrical resistance ratio between anodes and cathodes will also affect the voltage that the individual electrodes will propagate (see Section 4.6.2 on Kirchoff's Law).

In alternating current (AC) systems, both electrodes have the same charge characteristics (each alternately positive, then negative) and are usually the same geometric shape. In direct current (DC) and pulsed DC (pDC) systems, one electrode (anode) is positively charged and the other electrode (cathode) is negatively charged. Positive and negative electrical fields have different effects on fish, and thus the two electrodes differ markedly in shape, reflecting the different electrical properties and roles that they have to perform. The optimal characteristics of an electrode system were summarised by Novotney and Priegel (1974):

- To provide the largest region of effective current gradient in the water
- To minimise areas of damaging current density

- To be adjustable to cope with differing water conductivity
- To be manoeuvrable around weed beds and other obstructions
- To allow visual observation of the fish and thus enable capture.

Construction material should be lightweight (for hand-held anodes) high-conductivity metal; dirty or corroded metal will result in higher resistance values and thus affect the field gradient from the electrodes (Beaumont *et al.* 2002). For this reason, stainless steel is commonly used for anode construction. Aluminium has considerable advantages regarding weight, but if used it must be kept clean of the insulating oxide layer that rapidly builds up (the same applies to copper tube anodes). Titanium alloy electrodes provide low-weight, minimal corrosion and low electrical resistance but at a high cost. Copper or stainless-steel braid is often used for cathode construction due to its flexibility and the ease with which the long lengths required for optimal anode–cathode resistance ratios can be transported.

5.3.1 Anodes

The term 'anode' is commonly considered to include both the positive electrode and the system by which it is held and used by the operator. The design of this equipment has changed dramatically over the years – although some very poor early designs are still being used at the present time.

Early anodes rarely had safety switches on them, and if operators fell into the water there was a very real risk of them receiving an electrical shock before the system could be shut down at the generator – assuming that there was some-one stationed by the generator! Construction material of the anode pole was commonly wood (often wooden broomsticks), and as long as they were coated in a good layer of varnish these were adequate. However, in time, the varnish cracked or wore thin and allowed water to soak into the outer layer of the pole; this ultimately allowed the electricity to track up the pole to the operator. Most modern legislation, where it exists, now prohibits the use of wood in anode construction.

In recent years, dead-man's switches have been incorporated into the anodes; these require a constant pressure on a switch to energise the electrode. Early switches on pDC output systems directly switched the main 'fishing' voltage (Williams 1984). This could lead to the switches burning out or arcing across contacts. For this reason, modern switches are low-voltage (<24 V) units that switch a solenoid relay back in the pulse box. Most systems use single-pole switches, but some authorities (e.g. UK Environment Agency) specify that double-pole isolating switches should be used. The difficulty of maintaining continual pressure on a switch whilst holding the anode has led to some use of 'third-party' switching of the anode. One such system is based on one (or more) of the fishing team holding a radio transmitter that when active allows the anode to operate. It is important that this type of system 'fails to safe' (i.e. the relay switch *must* only operate if it gets a valid radio signal). Another system relies on a separate switch on a long wire that is operated by a person on the bank (or in the river). I have observed operators of the latter system throwing the anode upstream in order to capture shoals of fish moving ahead of the fishing team in the fast water that they were fishing, a method that would not be possible with an anode-mounted switch. The team using this system were experienced and had a good safety system for when to energise the anode; as a result, I considered the method still allowed a 'safe system of work' to be achieved.

Dead-man's switches should be designed or shielded so that dropping the anode cannot inadvertently operate them. On some manufacturers' anodes, the anode switch can also control the pulse waveform being used. For example, on one of the Hans Grassle units, the anode can switch between DC and pDC, with the idea that operators can fish using the low-power-demand pDC and then switch to the low-injury DC when a fish appears. On the Smith-Root LR-24 backpack, the switch can be used to alter pulse modes, with the same principle of power saving while finding the fish. Although this ability to quickly alter the output seems an advantage, in the real world it is unlikely that single fish come to the anode in an orderly manner, and keeping track of which waveform you are using would also be difficult – did you press the switch once or twice?

Some manufacturers have incorporated immersion sensors in the anode or use electrical current sensing circuitry, so they cannot be energised unless the anode head is in the water. Others have displays indicating the output on them (although these tend to make the anode very heavy).

One thing an anode should not have on it is a catching net. Voltage gradients in close proximity to the anode ring are very high, and capturing a fish and holding it in this zone while the anode is removed from the water are very damaging to the fish. From a normally energised (triangular) anode, I have measured voltage gradients of nearly 7 $V.cm^{-1}$ in the middle of the anode. Transferring the fish from the anode net to a holding bucket also will necessitate a bucket carrier to come into close proximity with an anode that, whilst (hopefully) not energised, would be energised if the dead-man's switch was accidentally operated or had failed. When fish are netted (using conventional hand nets) from the water, they should immediately recover once they are removed from the electrified water. I regularly see photographs of fish lying incapacitated on a shallow net that is attached to an anode; those fish should be so active that it is impossible to keep them on the net. The fact that it is immobile indicates that it has been over-shocked and may have suffered injury. In Europe, legislation prohibits the use of nets on anodes, stating 'nothing shall be taken from the electrode by hand' (Anon. 2003a) – although they are still widely used.

In normal electric fishing operations, the anode is potentially the most dangerous piece of equipment used. Operators are at greatest risk if a live anode is held out of the water where operators may come into contact with it. For this reason, anodes **must** only be energised when in the water.

Boom-mounted anode arrays are so called because the anode head is mounted on a boom fixed to a boat and not hand-held. They are usually controlled by foot-actuated relay switches, often in conjunction with pressure-sensitive mats on which the net operators stand.

Electrodes can also be constructed for use by divers (James *et al.* 1987). The hand-held probe is, obviously, waterproof apart from the two electrodes (anode and cathode) that are mounted 15 cm apart at the distal end of the probe. The diver, wearing a drysuit, places the probe as close as possible to the fish and activates it. Fish within 20 cm of the electrodes were immobilised with no ill effect on the diver.

5.3.1.1 Anode shape

The role of the anode in DC and pDC electric fishing is to create a zone near to or at the water surface, to which fish are attracted. The ideal shape is a large metal sphere; this gives an evenly distributed voltage gradient in all dimensions around it without any high gradients caused by corners and so on. However, this shape is unwieldy when used as a hand-held anode, and in classical (wading) electric fishing one shape (the ring or torus) predominates. The principle of the ring configuration is that it produces a horizontal electric field similar to the shape produced from the optimal spherical electrode shape, but it is a much more practical shape for use while wading. Other anode shapes are in use with various organisations, predominantly square, triangular or flat plates of expanded mesh (Figure 5.8).

Figure 5.9 compares the voltage gradient patterns from various shapes of anode. The anode field is of course three-dimensional, and voltage and power gradients will be present in the anterior, lateral and vertical axes to the anode.

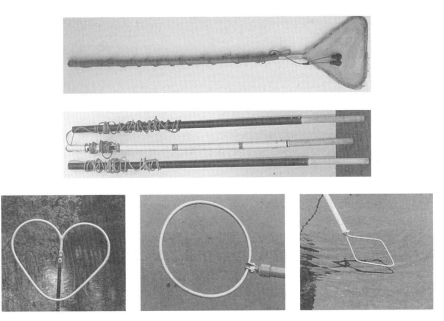

Figure 5.8 Various anode shapes in use: note the net on the triangular anode.

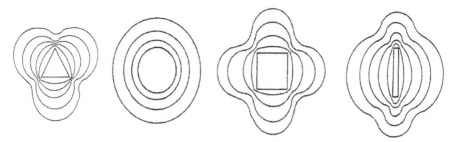

Figure 5.9 Schematic diagrams of voltage gradient patterns from differing ring, square and cylinder anodes.

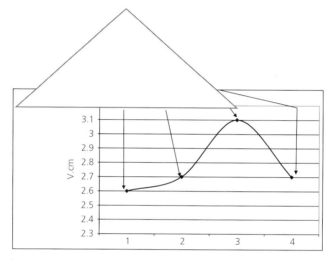

Figure 5.10 The measured voltage gradient around a triangular anode.

All anode shapes have their own characteristics, but generally a circular ring is considered the most efficient and practical design. Designs that incorporate acute angles or points are not recommended. If the triangle shape in Figure 5.9 is taken as an example, if the gradients projected from the flats of the square are sufficient to produce narcosis, the corners will have higher field gradients around that angle and may produce damaging tetanus. Figure 5.10 shows the measured voltage gradient from a triangular anode and the high corner gradient produced. These corners of the anode are commonly used to 'lift' fish from the river bed, therefore exacerbating the damaging effect of using this shape. Similarly, the tubular or cylinder anode in Figure 5.9 will have potentially damaging gradients at the top and bottom, and measurements taken of voltage gradients propagated from tubular electrodes show high voltages close to the anode. Tubular electrodes have been used predominantly in deep-water fishing from small boats, where they often have a dual role of both incapacitating the fish and manoeuvring the boat; unfortunately, they perform neither function well.

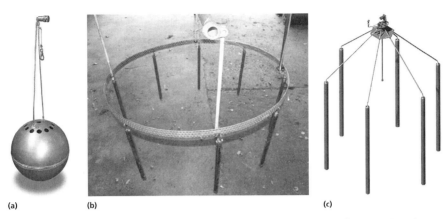

(a)   (b)   (c)

Figure 5.11 (a) Sphere anodes; and (b and c) Wisconsin ring anodes. (With permission of Smith-Root Inc. for photographs a and c.)

When fishing from boats, anodes can be mounted on booms attached to the boat (creating boom-boat electrofishers). On these, two designs of anode predominate; spheres (Figure 5.11a) and 'Wisconsin rings' (Figure 5.11b and 5.11c). When boom-mounted, the sphere does not have the handling disadvantages of when hand-held and, as stated earlier, projects an 'ideal' spherical voltage gradient pattern. However, in order to project large-diameter gradient fields (without having high near-anode gradients), the size of the sphere needs to be very large making its use less practical. To overcome this problem, an anode array called a Wisconsin ring (Figure 5.11b and 5.11c) can be used (Novotny & Priegel 1974).

In Wisconsin rings, the anode consists of a collection of vertical 'droppers' either suspended from a metal ring (Figure 5.11b) or without the supporting ring (Figure 5.11c) arranged in a circle. The principle behind the array is that the droppers are close enough that the voltage field around them joins and mimics the electrical field pattern of one large ring, but without the ergonomic and practical problems of actually using one. The droppers do not get caught up on weed and can hang down to a variable depth in the water, thus increasing the depth at which the electric field is effective.

5.3.1.2 Anode size

Anode diameter is one of the most important factors affecting the size of an electric field in water (the other being applied voltage). Two parameters govern the optimal diameter of the electrode: power available to energise the anode (and cathode), and physical stream size and topography (you cannot fit a large-diameter ring in a stream where many rocks and boulders project above the surface). The basic rule regarding anode size is 'use as large a diameter as possible' as large anodes will result in increased efficiency (capture field) and reduced fish mortality (Bohlin 1989). Within this ideal, restrictions will be based upon having sufficient power to energise the anodes and the physical size constraints noted above. Power supply requirements can be calculated either from theoretical electrode resistance or from empirically measuring resistance for particular electrode set-ups and then calculating power required. If power supply becomes limiting (e.g. in high-conductivity water), a combination of reducing all or some of a choice of ring size, waveform (DC vs. pDC), pulse frequency or voltage can be done to reduce overall power requirements.

The electrical field around an electrode should ideally attract and immobilise fish from the greatest possible distance, with optimum use of energy and least damage to fish (Davidson 1984). Key to the production of such an idealised anode is the knowledge of the field gradient produced from the design at different voltages.

Figure 5.12a–c shows diagrammatically the relationship between the radius of a torus-shaped electrode and the resultant sizes of the danger and effective zones propagated. This shows the way that anode size and energising voltage can be optimised to produce a good, non-injurious anode field. However, in the 'real world', there is always likely to be a zone of high-intensity voltage gradient near to the anode, and every effort should be made to capture the fish before they reach this zone.

Measurements of actual voltage gradients from a commonly used range of anodes are shown in Figure 4.10 in Section 4.1.1.

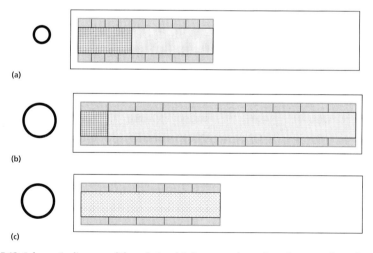

Figure 5.12 Schematic diagram of the relationship between the radius of a torus-shaped electrode, applied voltage and the resultant sizes of the danger and effective zones. (a) Electrode of radius r; electrode potential x V; (b) electrode of radius $2r$; electrode potential x V; and (c) electrode of radius $2r$; electrode potential $x \div 2$ V. Rings represent electrode sizes, vertical gridlines are equivalent to the electrode diameter, red hatching represents the danger zone and green hatching represents the effective zone

Table 5.1 Voltage gradients around a partially submerged anode.

Percent submerged	Voltage gradient (V.cm⁻¹) at 60 mm
100%	0.95 V
50	0.90 V
25	1.45 V

As previously stated, you cannot get a large anode into a small stream. In these circumstances, a smaller anode should be used, and the temptation to only use part of the ring in the stream should be resisted. When the anode is only partly submerged, high-voltage gradients can occur and thus increased injury is possible. Beaumont *et al.* (2002) took voltage gradient measurements from partly submerged anodes and found that when an anode ring was only 25% submerged, gradients at 60 cm were 150% of when fully immersed (Table 5.1).

Similarly, if using smaller anodes, voltage to the anode will need to be reduced in order not to create high-voltage gradients near the anode.

Alternatively, metal mesh can also be added to the anode in order to modify the voltage gradient, the effect of this being to extend the voltage gradient further from the anode. Under no circumstances, however, should the mesh covering be used as a net or used to lift fish from the water due to the risk of damage to the fish from high-voltage gradients at the mesh surface.

5.3.1.3 Twin and multiple anodes

The range to a 0.1 V.cm⁻¹ gradient (the edge of the pDC capture field in moderate-conductivity water) from a 400 mm anode with a 3 m braid cathode and energised at 200 V is about 1.5 m. If fishing a wide river (wider than ~5 m) with a single anode (at the output settings discussed above), fishing efficiency is likely to be reduced due to fish having space to escape around the electric field. For example, single-anode capture efficiency for salmon parr on the river Frome in Dorset (United Kingdom) reduced from an average of 72% on a 4–5 m wide carrier to 56% on the (~10 m wide) main river (author's data). Due to the differences in how circuit resistance between series (single-anode) and parallel (double-anode) circuits are calculated, however, it is far preferable to increase the number of anodes rather than increase the voltage of just one anode. If a single anode had its power increased to the level that would fish the area that twin anodes could fish, then the power requirement would dramatically increase and result in dangerously high fields close to the anode. Cuinat (1967) noted that if the number of anodes is doubled, then the power required is also doubled (assuming minimal cathode resistance) due to the change in total resistance of the anodes. For example, a single 400 mm anode and 3 m cathode in 400 µS. cm⁻¹ water and a circuit voltage of 200 V will give a 0.1 V.cm gradient at 1.5 m and have a peak power demand of 980 W. Doubling the number of electrodes

Table 5.2 The effect of adding additional anodes and cathodes on the voltage projected ring from the anodes (anode 26 Ω, cathode 20 Ω).

Anode diameter	Cathode size (copper braid)	Circuit volts (Vt)	% Circuit volts at anode	Anode voltage (Va)
400	3000	200	57	114
2 × 400	3000	200	40	80
2 × 400	2 × 3000	200	57	114

(anodes and cathodes to keep the resistance ratio constant) would use a peak power demand of 1970 W. Increasing the single-anode voltage such that the distance to 0.1 V.cm doubled to 3 m (500 V) would require 6150 W (peak). Increasing the number of anodes is therefore more efficient regarding power usage, provided care is taken not to overload the pulse box or the generator. Guidelines given in Northern Ireland (Kennedy & Strange 1981) suggest that one anode can deal with a river of up to 5 m wide, two anodes up to 10 m, three anodes up to 15 m and so on.

From the section on Kirchoff's Law (see Section 4.6.2), it can be seen that adding more anodes changes the total anode equivalent resistance. Cuinat (1967) estimated that resistance increased by 1.6 when comparing his two-anode and one-anode systems. Cathode equivalent resistance should there-fore also be changed to maintain the anode–cathode resistance ratio and thus the anode voltage (Table 5.2). Therefore, if multiple anodes are used (in order to increase the area fished), the area of the cathodes should be increased appropriately. If this is not done, lower anode voltages, which are counterpro-ductive to increasing the area fished, will occur. In the United Kingdom, the newer twin-anode pulse boxes are being fitted with facilities for twin cathodes; however, few of the older pulse boxes have extra sockets for additional cathodes. In these cases, twin cathodes can be wired into a single plug and the cathodes separated in the water (if they are in close proximity, the electrical advantage is reduced).

When twin anodes are fished in close proximity, the current draw on the circuit is likely to be increased due to the reducing resistivity of the electrodes when they are in close proximity. Care needs to be taken, therefore, not to over-load pulse boxes or generators.

In addition, the voltage gradient profile (Figure 5.13) of each anode is also altered (Beaumont *et al.* 2002). At a separation of 5 m (the river width at which the use of two anodes should be considered), there was little effect on the gradient field when using twin 600 mm anodes; however, when fished at 2.5 m separation, a considerable effect on the anode field is apparent. Therefore, when fishing twin anodes, care needs to be taken to keep them sufficiently apart that they do not interact with each other and reduce the

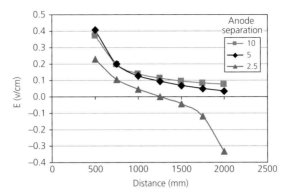

Figure 5.13 The effect of inter-anode separation distance (m) of twin 400 mm anodes on the voltage gradient from each anode. Circuit voltage = 200 V; anode voltage = 85 V.

voltage field they each project. Fishing twin anodes in the manner recommended in Chapter 7 ('Working techniques') will minimise the effect on the voltage fields.

5.3.1.4 Anode ergonomics

Although not as important as electrode head geometry, the ergonomic design of hand-held anodes can have an effect upon the ease of use of the electrode. Using an unwieldy and cumbersome anode will increase strain and the likelihood of injury to operators. In the past, little option was given as to the designs available, almost all being a simple straight rod with a ring attached either in line with the rod or angled so it was in front of the rod (Figure 5.14a). Holding the anode necessitates the operator having to flex and extend their wrist and results in considerable strain on the wrist. Where the ring is angled, the design results in a turning momentum on the handle (when facing upstream, the water flow constantly tries to rotate the anode head to a downstream position relative to the handle). New anode designs have recently been produced that allow a more natural grip (without wrist flexing) on the electrode that overcomes these problems (Figure 5.14b–d). Designs such as those shown in Figure 5.14d also allow a more natural and stronger grip as the thumb is used to grip the handle, as opposed to only the fingers gripping the handle (a 'monkey grip') and the thumb operating the dead-man's switch (Figure 5.14b and 5.14c). Improved material design also means that most electrodes can be light but still retain their robustness. As a general rule, maximum anode weight should be around 1.5 kg.

Electrode handles should be made from nonconducting material. Wood is not recommended and is banned in Europe (Anon. 2003a), as the outer layer can become waterlogged and allows electricity to track up the pole to the operator.

Figure 5.14 Examples of different anode designs: (a) Straight handle anode supplied by Electracatch International Ltd (UK); note switch has safety cover removed. (b) Ergonomic anode (thumb-operated switch) supplied by Intelysis Ltd (UK). (c) Ergonomic anode (thumb-operated switch) supplied by E-fish (UK) Ltd. (d) Ergonomic anode (finger-operated switch) supplied by a Hans Grassl Inc. (Germany) dead-man's switch can also control whether DC or pDC is output from the anode.

5.3.2 Pre-positioned area samplers (PPAS)

Standard electric fishing methods (wading in the water using a mobile anode) can disturb fish from their habitat prior to capture (due to a fright response both from the edge of the electric field and from disturbance from the operators wading in the river) and can also attract fish from one habitat to the habitat where the anode is positioned (due to anodic attraction). This can make assessing fish communities from discrete (microhabitat) areas difficult.

One electrode design that has been utilized to limit disturbance is the PPAS equipment described by Bain *et al.* (1985), Weddle and Kessler (1993), Fisher and Brown (1993), Baras (1995) and Schwartz and Herricks (2004). These arrays only incapacitate fish in a discrete area. Areas sampled will typically be about 1 m² in order that discrete habitats can be assessed. Electrodes can either be wire or metal tube laid out on the streambed or square frames with electrodes on opposing sides of the frame. Some arrays incorporate a catching net downstream of the array.

The PPAS is set in position some time (hours or even days) before energising to allow fish to recolonise the area. They are then energised and sample a discrete habitat or area without the fright bias of conventional gear. They are often powered by AC in order that fish are immobilised *in situ* (i.e. no electrotaxis to the electrodes) and often use high-voltage gradients (>5 V.cm) to kill the fish.

5.3.3 Point abundance sampling using electricity (PASE)

Another type of electrode arrangement that is designed to sample small, discrete areas is the point sampling gear (Copp 1989). The equipment is predominantly used for sampling juvenile fish from littoral areas.

The cathode is of standard construction, but the anode consists of a long handle (often >2 m) with a small-diameter (5 cm) ring on the end (Figure 5.15). The long handle minimises visual disturbance from the operators, and the small anode ring gives a small, high-intensity gradient to instantly immobilise the fry. The method used is to select a sample point (often by randomised sampling strategy) and then stealthily, but quickly, immerse and energise the anode at the point and then quickly sweep a hand net (with suitably fine mesh) through the area. The netted area and volume are estimated, and multiple points are

Figure 5.15 PASE anode being used to fish littoral margins from a boat.

sampled. A density estimate based on the mean fish per area caught and the area available can then be calculated.

Janac and Jurajda (2005) found that PASE (called by them 'Remote Electric Fishing') was 30% less efficient than the PPAS methods described here, but the relative abundance and size structure of fish assemblages were similar.

5.3.4 Electric nets

Electric 'nets' or 'seines' do not necessarily incorporate true 'nets' into their design. They are often just an array of electrodes suspended from ropes that can be moved up and down a stream for driving and incapacitating fish that are then collected by traditional nets (Sternin *et al*. 1976). The first mention of an electric 'seine' appears to be by Funk (1949) where a 90–100-foot-long device was used, powered by AC at 100 V. It was considered that the device was highly effective in sampling large rivers that could not otherwise be sampled. Despite the success of Funk's equipment, only limited use has been made of electric seine nets in freshwater, Bayley *et al*. (1989) found that only three of 18 groups of North American researchers were using the method. Bayley *et al*. (1989) also found that the method was more effective for estimating species number and abundance than standard (net) minnow seines or pDC backpack fishing. Lui *et al*. (1990) increased the catch rate of carp from a lake with an uneven bottom from 5% to 20–30% by using an AC-powered bottom 'foot-rope'. This foot-rope prevented the carp from escaping the net by means of burrowing under it, a common way in which they previously avoided capture.

Cave (1990) used a trammel net fitted with dropper electrodes powered with DC to attract and capture adult Atlantic salmon. However, no data on injury rates or capture mortality are given.

Despite the huge electrical current required due to water conductivity (often thousands of amps), electrical devices have been used in marine fisheries. The most common of these are electrical 'tickler' chains that are used in demersal marine trawl nets to lift fish and shrimp off the bottom so they will go into the net. This method is particularly effective where muddy substrate would otherwise make trawling difficult and also allows shrimp fishing in clear water and during daylight, when shrimp are inaccessible to conventional fishing (Monbiot 2015). Sternin *et al*. (1976) describes several types of electric fishing equipment used in large freshwater and marine demersal and pelagic fisheries. Electrical fields were used with pair trawls (where catch increased by 300%), in sardine liftnets where an increase of 80% was found and also to incapacitate fish that had been attracted by lights to 'fish-pumps'. The increased efficiency of the net-based systems allowed operators to increase the mesh size of their nets and retain catch weight whilst reducing catch of juveniles, and thus fish in a more sustainable manner.

In order to reduce the 50% loss rate from longline fishing, electricity can also be used to incapacitate large marine fish (e.g. tuna) when they have taken a baited hook (a method described in Baggs' patent in 1863) and also to kill whales that have been harpooned (Sternin *et al.* 1976).

5.3.5 Cathodes

The importance of the cathode in DC and pDC fishing is often overlooked. Whilst it is common for descriptions of electric gear to give anode dimensions (although not as common as it should be), it is rare for descriptions to include details of the cathode used. If it is too small, an intense field will be produced around it, which will adversely affect the fish. In addition, the reduction in voltage at the anode (from Kirchoff's Law) will necessitate increasing the anode voltage to maintain capture area and therefore necessitate the use of a larger than needed generator. Variations in the size and condition of cathodes between teams could also result in variations of catch efficiency between those teams and lead to lack of standardisation of results between teams.

Generally, the size of the cathode should be as large as possible; this will reduce its electrical resistance, lessen the voltage gradient around it, allow more of the circuit voltage to be apportioned to the anode and make more efficient use of power. Vincent (1971) recommended an anode–cathode area ratio of 1:30 as being sufficient. However, there is a law of diminishing returns between increasing the cathode size and its electrical resistance. Cuinat (1967) found that if the size of the cathode was doubled, its resistance halved, and recommended an 80 × 60 cm panel of 2 cm mesh as being sufficient. Beaumont *et al.* (2005) found a power law relationship between 25 mm width copper braid cathodes and equivalent resistance, with the relationship flattening out at around 3000 mm of braid.

In the past (pre-1980s), the usual construction for cathodes was a sheet of expanded metal, but since then there has been an almost universal switch to using lengths of copper or stainless-steel braid or, more recently, steel wire. Several authors comment that a cathode is more effective when its form is least concentrated; in this respect, the braid construction should result in a highly effective surface area, thus resulting in a low field intensity and low power loss. To test the assumption that the braid is more suitable for cathode construction, Beaumont *et al.* (2005) took measurements of the voltage gradient around different designs of cathode. The results indicated that the electrical resistance of the commonly used braid design is not markedly dissimilar to that of a plain metal tube and also that the difference between a metal mesh sheet and solid plate metal was also minimal. However, the braid and wire cathodes have several ergonomic advantages in terms of ease of transport and use over the metal tubes or sheets.

Some users advocate the use of several small cathodes in preference to one large one (this would put the cathodes in a parallel resistance configuration and so reduce their overall electrical resistance). If multiple short cathodes are used, they should be widely separated; otherwise, they will be electrically coupled and resistance will not be as low as predicted.

Practical considerations, especially when towing equipment in boats and the like, may make single or possibly double lengths of braid the first choice due to problems of weed snagging and for storing and transporting equipment easily. However, copper braid does fray badly when it is pulled through rocky rivers behind fishing units; in these cases, the use of twisted-steel wire (of about 5–6 mm diameter) is preferable as it is extremely hard wearing and also less likely to snag on or between rocks and obstructions in the river. For backpack equipment, it is less practical to have long (3 m) cathodes due to the need for them to trail through the river behind the operator (problems with colleagues stepping on the cathode being just one of the issues). Backpack cathodes are therefore often shorter than those used on fixed-position equipment.

If the substratum of a stream is very conductive, it can have a profound effect upon the equivalent resistance and thus the power drawn by the cathode. Under these circumstances, it may be necessary to use a floating cathode that does not come into contact with the streambed (e.g. Figure 5.16).

The particular design of floating cathode illustrated in Figure 5.16a is not recommended due to the conductive upper surface. This does not contribute to the in-water area of the cathode (and thus its equivalent resistance) yet poses a safety risk (the attachment point in the centre of the disc also resulted in the disc tipping and forming a drag anchor!). Figure 5.16b uses a sealed plastic pipe as the floatation device with a braid cathode rigged underneath; this gives good towing characteristics and keeps the cathode underwater.

It should be remembered that if fishing multiple anodes, cathode area or number should also be increased. Failure to increase the cathode area could result

 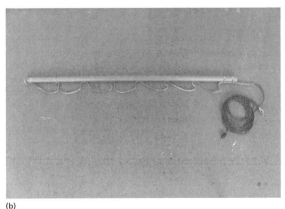

(a)   (b)

Figure 5.16 Examples of floating cathodes.

in lower anode voltages being produced (from Kirchoff's Law) and thus capture area reduced from each anode with resulting reduction in fishing efficiency.

Even when using a 3 m length of braid, the cathode may produce an incapacitating electric field. Placing the cathode in fast-flowing water will prevent fish from being held in the field and being injured. If the cathode is placed in shallow water at the upstream end of the fishing reach, it can also act as a partial barrier to fish movement (due to the repulsing effect of a cathodic voltage gradient) and thus increase capture rates and possibly reduce the need for stop nets.

Cables on cathodes should be long enough that the generator position (level surface away from flammable materials) is not compromised by the need for good cathode placement.

5.4 Hand nets

Many commercially available nets are designed for heavy-duty fish farm use and are therefore of very substantial construction, often requiring two hands to manoeuvre and control them (Figure 5.17). Net aperture size on the commercial nets can also be very large. When netting incapacitated fish, however, net aperture

(a)   (b)

Figure 5.17 (a) A commercial hand net; and (b) in use during electric fishing operations. Note the close proximity of the metal to the operator's hand.

can be small and provided the head of the fish is targeted for capture, even large fish can be caught with small nets (the author has caught 10 kg+ Atlantic salmon and pike in a net with an aperture measuring just 250 mm × 250 mm).

The large amount of metal in the construction of the commercially available nets will also pose a safety hazard (making it possible for hands to touch the metal part of the net handle whilst the net is in the electric field). If gear is set up to just incapacitate the fish, the introduction of a large metal net may also alter the electric field sufficiently for the fish to no longer be incapacitated (author's observation). For these reasons, the amount of metal in the net should be minimal, and any present should be insulated: this is easily achieved by simply wrapping the metal in heavy-duty insulation tape. Nonconductive material for making netheads would be ideal, but it would need to be able to be strong yet not too cumbersome.

When electric fishing, it is often advantageous for one person both to control the anode and to net fish. This gives a reduced manpower benefit and can often result in more efficient fish capture (especially in narrow streams). If one operator is both controlling the anode and netting, the nets need to be designed, so that they can be used one-handed without causing strain to the operator (Figure 5.18 and 5.19).

Figure 5.18 Using a lightweight net in one hand and an ergonomically designed anode in the other.

Figure 5.19 Lightweight net designed for use from boats. Net is 2.2 m long, has an aluminium head and weighs just 900 g.

Figure 5.20 A 'banner net' and one being deployed in the river (with the anode fishing downstream towards it).

Handles of nets should be made of nonconducting material (Anon. 2003a). If wood is used, they should be suitably varnished or waterproofed (and maintained in that condition) so that 'wicking' of the outer layer does not occur, thus allowing the outer layer to become conductive. Some use of bamboo has been tested (this wood does not become waterlogged in the same way that other woods do); whilst it has shown promise, when it does split the sections become filled with water and thus unusable.

5.4.1 Banner nets

These nets form, in effect, a mini static seine net. Two (insulating) poles have a curtain of mesh suspended from them (Figure 5.20). The net operator holds the net in flowing water downstream of the anode, and the incapacitated fish are carried by the water current into the net. The net is then lifted, and the fish are transferred into the holding bin. The net requires two hands to operate, so the operator can only do this task. Care needs to be taken that the operator does not allow both hands to touch the water and create a circuit; for this reason, insulating gloves are recommended if working in moderately deep water where that risk exists. The method works best when fishing downstream and works particularly well in high-velocity, steep-gradient streams where 'normal' hand net use could be difficult. With experienced operators, the method is very efficient and quick.

5.5 Stop nets

When requiring population data from discrete areas of river, some method of isolating that reach (i.e. stopping the fish from swimming out of that reach as a result of the general disturbance that takes place during fishing) is needed. The common way of doing this is to use relatively fine mesh (1–2 cm) 'stop nets' (sometimes called 'block nets') at the top and bottom of the reach.

In certain circumstances, stop nets are essential to determine accurate population estimates. Peterson *et al.* (2005) found that with single-pass electric fishing without stop nets, 18% of bull trout moved upstream out of the fishing reach; this percentage declined with increasing levels of 'rubble' substrate. However, stop nets are not always necessary; Simonson and Lyons (1995) found that stop nets had little effect on catch per effort during single-run upstream fishing; and when the length of stream fished was 35 times the width, the effect was negligible. Bohlin (1989) also considered that stop nets were not necessary in small streams and found that the longer the reach length, the less the 'edge effect' of disturbance affected fish moving out of the reach. Similarly, Edwards *et al.* (2003), in a comparison study of single-pass versus depletion surveys, found that stop nets were not necessary. As a *very* generalised rule, if you are fishing a relatively deep (>40 cm) reach with little fish cover or rocky riffles, then you are likely to need a stop net. This is particularly so if very mobile pelagic living fish (e.g. cyprinids) are present. However, if the reach has plenty of fish refugia or you are fishing in a shallow pool and riffle-type stream, particularly if more territorial fish (e.g. juvenile salmon and trout, or benthic living fish such as cottus spp.) are the target species, then nets may not be necessary. If the start and finish of a reach can be made to coincide with a shallow riffle and the cathode is placed at the upper riffle, then the electrical barrier so formed should also inhibit fish movement out of the reach. Deeper rivers where fishing from a boat is required will certainly need nets to prevent fish swimming out of the reach.

Skilful anode use (discontinuous energisation and good awareness of likely fish habitat) will also reduce the need for stop nets.

The high level of debate and interest in whether stop nets are necessary stems from the fact that they are very difficult to install properly, particularly in areas where drifting weed and debris are likely. Even if they are initially installed such that they are 'fish tight', drifting weed and debris will often render them ineffective in a short space of time. I have commonly seen various individuals and organisations fishing a reach with the 'used stop nets' box ticked, but with the net of the bottom stop net not on the bottom of the stream.

Due to the need to have the nets sit tight on the bed of the river, the bottom line of the net (leadline) should be heavily weighted. Weights need to be closely spaced so that the net does not go off the bottom between weights, and metal chain is very good for giving a continuous flexible weight along the bottom of the net. The net should also have good floats on the top line or have the top of the net suspended above the water line.

When setting the nets, operators should create the minimum disturbance possible, particularly if the reach length is short. Setting the nets should also be carried out before any other activities take place on the river (to avoid scaring the fish out of the section).

The nets should be set in slow-flowing water and set at a shallow angle to the vertical, with the lead-line upstream so drifting debris and fish are guided to the top of the net. If anchored using metal stakes, care needs to be taken regarding buried power cables or pipelines. The most common error with stop nets is thinking they are set, only to find that drifting weed and debris (for the downstream net, often disturbed by the operators walking in the stream) have collected in the net and caused it to either lift off the bottom of the stream or sink the float line, thus allowing fish to escape the section. For this reason, some operators use either heavy weights that are placed in the centre section of the net or tripod-type arrangements that both hold the bottom of the net to the streambed and keep the float line at the surface.

A variation on the 'standard' stop net is the electrified barrier as used by Mann and Penczak (1984). In this study, fishing was carried out using pDC (220 V, 50 Hz) from three boats drifting downstream to an electrified AC (220 V) barrier spanning a shallow section of the river (with one electrode on the bottom and one just sub-surface). Fish passing the AC 'stop net' were killed and were netted or noted by netters stationed ~25 m downstream of the 'stop net'. The AC barrier was used as traditional stop nets were considered impractical (possibly due to the river width of between 50 and 90 m). Capture efficiencies of between 28% and 82% (depending on species) were achieved (as determined from Zippin estimates) with ~11% of all fish captured below the AC 'stop net'.

5.6 Protective and safety equipment

When working in wet environments using live electricity, it is vital that good personal protective equipment (PPE) is available.

5.6.1 Waders

It goes without saying that good-quality waders should be worn when wading in water surrounded by an electric field! These should be checked for holes and tears before entering the water, and spares or a repair kit carried. Large tears in waders will usually require patches to reinforce the area, but small tears and punctures can be repaired using gel-type glues. Repairs often need several hours for the glue to cure and set; however, there are some repair fluids that set in sunlight in just five minutes, and these are good for quick repairs in the field.

For safety, and particularly in fast-flowing water, wading should not be carried out in depths where there is a risk of operators losing their balance or going in over the top of their waders. As a general rule, thigh-length boots can be worn in knee-depth water, and chest-high waders in thigh-depth water. Wading in water deeper than thigh depth is dangerous (even if the water is slow flowing), as operators will start to become buoyant: it is also extremely difficult to fish efficiently while wading in such depths.

Certain types of rocky streams can be extremely slippery, and metal studs in the boot sole are excellent for increased grip. Care needs to be taken that these do not go all the way through the boot and thus enable electrical contact with the wearer's feet.

Breathable fabrics are available for wader construction, particularly for chest-high waders. These are advertised as allowing perspiration to leave the waders without allowing water in. When working in warm conditions this sounds very appealing, but some operators using these boots have noted receiving electric shocks through the material, particularly if wearing short trousers so that bare legs are touching the fabric. For this reason, it is advised (unfortunately) that, particularly if using high outputs, these are not used.

Similarly, whilst drysuits that have metal diagonal-entry or 'comfort' zips that are below water have not been known to transmit the electrical field to the wearer, for peace of mind it is preferable to use drysuits with shoulder-entry zips and no additional zip openings.

5.6.2 Gloves

In certain countries (particularly the United States), the wearing of electrically insulating 'linesmen's' gloves is mandatory. This is to reduce the risk of operators receiving electric shocks from faulty equipment, contact with the electrified water or inadvertent contact with live electrodes.

My experience of operators' views in Europe is that gloves are uncomfortable to wear, make using equipment (e.g. anode switches) cumbersome and impart a false sense of security to the wearer. In use, the inside often becomes wet, and then it is difficult to tell if the wetness is just from sweat, water entering the glove (due to routine net emptying) or a leaky glove. These problems result in operators being reluctant to wear gloves, even when mandatory. Hartley (1975) also considered gloves more of a hindrance than a safeguard.

Personal experience of using them leaves me unimpressed (particularly in hot and humid conditions) and concerned that they give a false sense of security that could lead to practices that would not be considered if the operator did not think they were 'protected' by the gloves.

My preference is for equipment to be designed such that it is intrinsically safe. It should have good and regular servicing. Well-trained staff with safe working practice protocols will also reduce any risk to acceptable levels. These measures are, I feel, preferable to using equipment or working practices that need gloves to allow safe operation.

5.6.3 Other protective clothing

Although electric fishing should not be carried out in the rain, good waterproof gear should be available for use in conditions where water from nets or spray may affect operator comfort, and thus efficiency.

Conversely, lightweight clothing that also reduces the risk of sunburn will be useful in sunnier climes.

Hats will also both protect operators from glare and reduce the effect of sun on their heads. Hats that can be dipped in the stream and worn wet are excellent for cooling when fishing under very hot conditions.

Sunglasses (with polarising lenses), although not strictly clothing, are a valuable aid to seeing fish in sunny conditions and reducing eyestrain in bright sunshine. Polarising lenses can also be incorporated into safety glasses and can be a safety aid if a net pole or overhanging vegetation comes into contact with them rather than directly with the wearer's eye.

5.6.4 Lifejackets

If operators fall into even shallow water while fishing and receive an electric shock, it is possible that they will be incapacitated and unable to swim. For this reason, it is advisable that self-inflating lifejackets are worn when electric fishing. When fishing in deep water or from a boat, then wearing lifejackets should be mandatory. Modern lifejackets are so small that little inconvenience should be encountered from wearing them.

CHAPTER 6

Practical factors affecting electric fishing efficiency

In addition to inefficient set-up of the electrical output settings and electrodes used in electric fishing, several other factors can also affect the efficiency of the technique.

6.1 Manpower requirements

Whilst it is important to have sufficient manpower to carry out electric fishing efficiently and safely, having too many people present can create problems with people getting in each other's way and compromising safety. W.G. Hartley's comments (personal communication) that "risk in electric fishing increases in direct proportion to the number of people involved" probably holds true for many situations, and excessive numbers of personnel should be discouraged and spectators kept well clear of operations.

Numbers of personnel needed for classical, wading, electric fishing vary depending on the stream width and number of anodes used. Most sampling on rivers less than 5 m wide can be carried out with three operators: one anode operator, who may also carry a net; one bucket carrier, who also may carry a net; and one person on the generator overseeing safety and gear settings. If the stream has significant weed or obstructions, an additional person may be needed to manage the cable from the pulse box to the anode. The use of 'tote boats' to carry the generator and pulse box can obviate the need for this extra person. If using backpack gear, then a minimum of two operators can be used provided all safety considerations are covered.

Although some types of net and trap systems require only one or two people, generally fishing using techniques other than electric fishing requires greater numbers than this. Electric fishing is also an efficient method of sampling in terms of catch per unit effort, with Pugh and Schramm (1998) finding that catch per unit effort can be higher for electric fishing than for other forms of sampling.

Electricity in Fish Research and Management: Theory and Practice, Second Edition. W.R.C. Beaumont.
© 2016 John Wiley & Sons, Ltd. Published 2016 by John Wiley & Sons, Ltd.

Boat-based fishing usually has a set number of personnel required to operate the boat and catch the fish; this can vary between two and five.

6.2 Streambed: conductivity and substrate type

As noted in Sections 5.1, 5.3 and 5.3.5, a very conductive streambed (saline sediment or metalwork) can 'short out' the anode and cathode; in addition, high-conductivity sediment can affect the electrode (particularly cathode) equivalent resistance. Scholten (2003) found a 20–30% reduction in fishing range over muddy substrate (compared with sand or gravel). He ascribed the cause of this as due to the current density line changing, which would point to the muddy bottom having higher conductivity than the sand or gravel. Localised changes in conductivity can also affect the efficiency during fishing (especially if using DC) and thus affect population estimate results. Sternin *et al.* (1976) considered that whilst rock or stone had a high resistance, the mineralised interstitial water between them was often of lower resistance. He considered that the ratio of bed to water conductivity usually fell in the range of 0.2–0.8, and rarely did the bed conductivity exceed that of the water.

Specific factors associated with substrate type include:
- *Rocks*: Streams that have large boulders or cobbles are difficult to fish due to fish either becoming uncatchable when narcotised or hiding under boulders.
- *Mud or silt*: This can affect subsequent fishing by reducing visibility. In certain circumstances, it can be highly conductive and can thus influence the current and power demand. In these circumstances, a floating cathode may be advisable. For sampling the lower tidal reaches of rivers, it is often possible to sample on the ebb tide when the saline water is flushed from the river by the freshwater flow. In these cases, it is important to realise that the sediment may still contain saline elements, and anodes and cathodes should not be allowed to come into contact with it. I have also heard of one river near a site of historic metalworking where there were so many metal filings and fragments in the river that it was impossible to electric fish.
- *Weed*: Problems can occur if the voltage (current) gradient is such that it immobilises fish within the weed bed instead of attracting them out of the weed. However, if the weed is very dense, fish may still become entangled and thus not be drawn from cover by the taxis effect of the fishing gear. In densely weeded streams, it may be very difficult to physically wade though the stream and impossible to obtain good quantitative data.
- *Metal pipe-work or bank reinforcing*: This can reduce the effect of electric fishing and potentially create a dangerous short-circuit of the anode and cathode.

Site topography, vegetation and turbidity can also affect fish capture efficiencies. Dewey (1992) found that whilst capture efficiency in non-vegetated, relatively clear water was around 80%, in vegetated turbid water it dropped to 5%.

Sammons and Bettoli (1999) found that catch rates varied due to different habitat uses by different sizes and species of black bass and that specific habitats contributed high variability to overall estimates of density (i.e. fish capture was less constant in those habitats).

Cunningham (1998) found that sampling flathead catfish was most effective where bank inclines were moderate to steep and bottom substrates were composed of riprap or natural rock, or where submerged structure was evident. It is not clear, however, whether sampling was truly more effective or whether these bottom types aggregated fish and thus just increased catches.

6.3 Weather

Electric fishing must not be carried out in conditions of rain, thunder or lightning. Wet equipment is likely to create conductive routes on the equipment with subsequent safety implications for operators. Conditions where lightning may be present are particularly hazardous, and gear should be shut down and stowed under cover or left lying on the ground in these conditions.

Heavy wind can also disrupt electric fishing due to waves on the water impeding visibility.

Bright sunshine can also affect visibility due to the glare off the water surface; under these conditions, polarising sunglasses should be worn.

6.4 Water temperature

Water temperature will affect the ambient conductivity of the water and thus the power demand of the fishing gear. Fish are isotherms (cold blooded); therefore, temperature will also affect their response to the electric field, due to suppressed physiological response times at low temperatures. The conductivity of the fish will also alter (in the same manner that the water conductivity changes); this may also affect the power transfer into the fish if the ratio of the fish and water conductivity is not constant. Lamarque (1967) noted that in cold conditions fish become immobilised less easily by electric fields, and he stated that the excitability of nerves decreased with declining temperature, recommending that at low temperatures the pulse width is increased in order to improve fish capture (Lamarque 1965). Both Justus (1994) and Cunningham (1998) found that sampling for flathead catfish was less effective at lower water temperatures. Borkholder and Parsons (2001) found that catch per unit effort peaked at 18.6 °C when catching 0-group walleyes (*Stizostedion vitreum*).

Behaviour at low temperatures may also prevent capture: Scruton and Gibson (1995) found that at low temperatures the fish burrow into river substratum, making them less catchable and also more vulnerable to damage from repeat shocking. They considered that salmonid fishing should not take place below 7°C due to these behavioural changes in the fish.

Conversely, in warmer water fish will become more difficult to catch due to their higher activity capability; Hayes and Baird (1994) noted that electric fishing became less efficient at higher temperatures. Vibert (1967a,b) found that at high temperatures there was less attraction of fish to the anode due to physiological reasons. Regis *et al.* (1981) also found that the 'effective attraction range' was lower in higher water temperatures. Differing fish species, however, will react differently and will also have differing threshold values to both cold and hot conditions. This threshold will almost certainly vary depending on acclimatisation, with the same species of fish in warmer climes having a higher tolerance than those from colder ones. The welfare of the fish post capture will also be affected by ambient temperature (see Section 13.7).

In summary, there is probably a temperature range over which fish react sufficiently to be caught by electric fishing. When too cold, higher gradients or longer pulse widths may improve capture rates (Lamarque 1967). When too hot, fish welfare post capture will probably preclude fishing. Present UK Environment Agency protocol suggests the best water temperatures for fishing to be 10–15°C for salmonids and 10–20°C for cyprinids.

6.5 Fish size

There is a general consensus among researchers that small fish are harder to catch than large fish (Snyder 1995). Early researchers often refer to the 'body voltage' needed to capture certain sized fish. Bary (1956) considered that the voltage gradient needed to elicit a reaction from the fish was related to the fish length (and two constants based on pulse duration). Kristiansen (1997) found a significant difference in the recapture probability between small and large sea trout and emphasised that size selectivity should be taken into account in electric fishing. Borgstroem and Skaala (1993) also found that fish length and catchability (with pDC backpack gear) of both juvenile trout and salmon were positively correlated in a low-conductivity stream. Some researchers have also found that the voltage required to kill fish is related to total head-to-tail voltages (Collins *et al.* 1954, Whaley *et al.* 1978). Lamarque (1990), however, noted that fish nerves have a maximum length of 4 cm; therefore, any effects should be confined to fish with nerve lengths shorter than this.

Some researchers have stated that long fish needed a lower pulse frequency than short fish to induce electrotaxis at the same potential. Halsband (1967) suggested that the reason for this was because big fish have big muscles and these big muscles are unable to relax between pulses if the frequency is too high. Fish size does seem to affect their susceptibility to injury from electric fishing, with large fish of the same species more likely to be injured (Sternin *et al.* 1976).

Dolan and Miranda (2003) studied the significance of a range of fish-size-related parameters on the power threshold levels (μW.cm^3) required for immobilisation. Fish volume was found to have the highest significance levels for power required, with a rapid decrease in power required up to a fish size of 50 cm^3; thereafter, the relationship became very weak. Similarly, Sternin *et al.* (1976) found that the maximum cross-sectional area of the fish (which would be related to fish volume) was very important in determining the response level.

Sternin *et al.* (1976) also considered that for the same length, older fish were more resistive (due to changing fat levels in the fish tissues) and therefore easier to catch (for a given voltage gradient).

Sampling small (<20 mm) juvenile fish (fry) is where it is acknowledged that high field intensities are required to immobilise the small fish. Electric fishing gear designed or modified for fry sampling (e.g. point abundance sampling) uses a small anode to achieve high gradients (Copp 1989). The technique is very effective, and Perrow *et al.* (1996) found that when used for qualitative and quantitative stock assessment of fry, point abundance sampling showed several distinct advantages over standard electric fishing within stop nets.

Pre-positioned area samplers (PPAS) also often use high-intensity gradients to create instantaneous and lengthy immobilisation of small fish (juveniles) in order that they can be collected from the area sampled.

The human bias in capturing different-sized fish should not be forgotten. It takes an experienced operator with good self-control to continue catching small fish when larger fish are also coming to the anode. Some bias is bound to occur, though, as larger fish are more affected by the electric field and will be attracted to the anode more readily than smaller or benthic fish. This bias is bound to affect population estimates and should be acknowledged.

6.6 Fish species

Some species are notoriously difficult to capture, and some very easy. To some extent, these differences may be attributable to the differences between fish and their conductivity (Section 4.8). Sternin *et al.* (1976) reported differing capture rates for same-length fish of different species and ascribed this to differing fish behaviours, body shapes and conductivity. Sternin *et al.* (1976) also considered that the nerve supply to the skin of different fish species affected their response, and quoted work by Bodrova and Krayukhin (1959) in which fish that had had their skin anaesthetised with Novocain required twice the voltage gradient to immobilise them. However, Dolan and Miranda (2003) found that power thresholds (from five differing waveforms) required to immobilise seven different species of fish remained constant, refuting any species-specific response.

With some fish, factors such as mobility, behaviour when startled and habitat being used when shocked will all influence capture efficiency. Very mobile pelagic fish (e.g. dace (*Leuciscus leuciscus*) and grayling (*Thymallus thymallus*)) are more able to avoid the electric field, particularly if an inexperienced operator is using the anode. Eels are noted as a difficult species to catch, and Lambert *et al.* (1994) noted that removal methods (for population estimation) were often not possible. Benthic fish can also be difficult to capture, possibly because once incapacitated, they remain on the bottom, making them difficult to see and collect (Cowx 1983). Their habit of burrowing into the riverbed may also enable them to withstand higher voltage thresholds due to the bed material acting as an insulating or conductive barrier (similar to a Faraday cage). Due to this burrowing behaviour, lamprey ammocetes are also difficult to capture. However, DC or low-frequency pDC (3 Hz) has been found to be effective for achieving good enough capture efficiency to carry out population estimation (Pajos & Weise 1994).

Seasonally or temporally changing habitat niches during the fish's life or at different development stages may also make the fish more or less vulnerable to capture. For example, if the fish migrate from deep areas of lakes or rivers to shallow or marginal habitats during spawning, they are likely to be more easily captured at this time.

Injury rates can also vary between species (Kocovsky *et al.* 1997), possibly due to differences in body conductivity. Salmonids are widely reported as being very susceptible to injury, but whether they are more susceptible than other species or have just been studied more often is not certain (Snyder 1995). Fish about to spawn may also be more susceptible to injury due to physiological changes in muscle and bone composition that may be taking place. In a survey of UK Environment Agency staff, dace (*Leuciscus leuciscus*), chub (*Leuciscus cephalus*) and grayling (*Thymallus thymallus*) were found to be the most sensitive species, and benthic species such as eel (*Anguilla anguilla*) the most robust (Beaumont *et al.* 2002).

6.7 Fish numbers

When electric fishing for population assessment data, it is important that not too many fish are incapacitated at any one time. This problem can lead to inefficient capture rates due to net saturation (i.e. more fish incapacitated in the water than can be caught in the available nets). This can affect both the physical number of fish caught (Zalewski 1983) and the welfare of the fish being caught. Even in moderate fish density streams, this effect can happen when fish are 'herded' by the fishing gear and congregate at the top stop net. Catch depletion sampling relies on the assumption that all fish are equally catchable; therefore, when the above happens, this assumption is likely to be invalid.

It is possible to reduce these problems by having experienced anode operators fishing in a discontinuous manner to reduce fish 'herding', and having them not move too swiftly up the stream. In addition, there should be sufficient net operators to capture the expected number of fish. If the situation arises of too many fish being incapacitated, then it is advisable to retreat downstream away from the shoal of fish and re-ascend at a much slower rate.

Where high numbers of fish are present, it is important that fish are not subjected to multiple exposure to the electric field by holding fish in nets whilst repeatedly netting more fish from the electric field. A system whereby nets can be quickly emptied into holding bins or full nets swapped for empty ones should be instigated. Care also needs to be taken where there are large numbers of fish present that adequate holding facilities are available for the fish, particularly if one must hold the fish during depletion sampling.

6.8 Water clarity

If the water is turbid, fish will not be seen even if affected by the electric field. Fishing technique can help to mitigate the problems that poor visibility brings. Using DC or pDC waveforms will attract the fish to the anode. Therefore, if the anode is kept high in the water (within view), fish attracted to it should be visible and thus catchable. Similarly, if the anode is drawn from deep water towards the operator, fish are likely to follow the anode and thus be caught (especially if using a strongly attractive current type). In general the electrode should be visible, and if quantitative sampling of all species is required, the electrode field should encompass the riverbed. However, if collecting benthic species, the streambed itself will need to be visible due to these fishes' poor electrotactic response.

6.9 Site length

It is important that an adequate length of river, ideally one that encompasses in proportion the entire range of habitat types present, should be fished. If only one reach of a stream is to be sampled, if it is to be considered representative of the stream as a whole, it should include the full range of habitat types present in the stream. Differing species may also require different reach sizes and sites in order to get a representative population assessment. Numerous small benthic species (e.g. cottus spp.) may only require sampling a few square metres; however, rare and/or mobile mid-water species (e.g. shad) may require hundreds of square metres to be fished.

Due to the financial implications of fishing multiple reaches or long sections of river, considerable interest has been shown in assessing the minimum reach length or number of sites that is required to sample to obtain consistent and cost-effective population data. Penczak (1985) found that the reach size required for minimum variation in population size estimation stabilized as the reach size increased, with reach sizes of 1125 m^2 being needed for some species. Hughes *et al.* (2002) concluded that a reach length of 85 times the wetted width would allow repeatable results and collect 95% of species present. If all species were required to be collected, then 300 times the channel width would need to be sampled. Vehanen *et al.* (2012) concluded that single-pass fishing required >450 m^2 to be fished to estimate fish assemblage attributes in small boreal streams.

6.10 Water depth

Pierce *et al.* (1985) found that at higher river stages, catch per unit effort was lower when sampling shoreline fish assemblages. However, they ascribe this difference to the fish being less abundant along the shoreline at higher river depths rather than the electric fishing being less efficient.

For safety reasons associated with the risks of drowning, present UK Environment Agency guidelines recommend that the maximum depth of water waded is hip deep, with the average being thigh depth.

Deeper water will require boat-mounted equipment, and the depth that can be fished will be dependent on the anode array being used. Hand-held anodes (400 mm diameter) are effective in water up to 2 m deep, but with reducing efficiency as the depth increases. Large (>1 m diameter) boom-mounted anodes will project a catching field much deeper but are still unlikely to be very efficient below a maximum of 3–4 metre deep water. Increasing anode voltages will give a larger catching field, but will also result in a larger, damaging, tetanising field. Bohlin (1989) considered that the catchability of fish gradually decreased with increasing depth, which then yielded a negative bias in fish population estimation at deeper sites.

6.11 Site width

A correctly set-up single anode of approximately 400 mm diameter should be able to effectively fish rivers up to 5 m wide (Kennedy & Strange 1981). Beyond that width, multiple anodes should be used in order to obtain an adequate capture efficiency if carrying out population assessments. If only single anodes are available, wide sites may be able to be divided by long nets into areas (lanes) that can be covered by a single anode (Figure 7.3).

When using multiple anodes, it is important that cathode size is also increased (see Section 4.6.2). The safety implications of using multiple anodes should also be assessed. The UK Environment Agency recommends a safety (dead-man's) switch arrangement that only energizes the anodes when *all* dead-man's switches are operated (a 'one off, all off' system).

Using multiple independent sets of electric fishing gear has serious safety implications due to the danger of gear not being shut down if an operator falls into the water. Strict codes of practice and highly trained operators are recommended if this method is used.

Mann and Penczak (1984) developed a new technique for fishing a very wide (50–90 m) river. The method used multiple boats and wading teams (each with their own independent power source) that fished down to an AC electrode array placed across the river. A further catching team was positioned downstream from the AC array.

For wide rivers and canals, however, boat-mounted, multi-anode boom-boats are probably the most efficient method of sampling (Cowx *et al.* 1990).

6.12 Time of day

Significant differences between day and night catch rates for a range of North American species have been found (Paragamian 1989, Dumont & Dennis 1997). Sanders (1992) found increases in taxa, numbers and total weight of fish caught (in large rivers) at night. This was attributed to fish moving inshore at night, where catchability was higher. Andrus (2000) attributed the significantly higher night-time catch in the main channel of the Willamette River (Oregon, USA) to the presence of day-time avian predators. These findings have led to it becoming common practice for electric fishing surveys (particularly boom-boat surveys on large rivers and lakes) to be carried out at night in the United States. However, Malvestuto and Sonski (1990) found that despite night fishing for bluegill resulting in a significant increase in catch rate compared with day fishing, proportional stock density estimates were still possible with just the day-time data. In the United States, although hydro-acoustic surveys have also revealed considerable differences between day-time and night-time patterns of fish assemblage, it is very rare for electric fishing to be carried out at night. Further work on the implication on time of day of fishing should be carried out.

CHAPTER 7

Electric fishing working techniques

Electric fishing is used for a wide range of fisheries management and research operations. These include fish rescue, broodstock capture, predator removal, stock assessment and research. Each of these has its own requirements for equipment set-up, but all should comply with the tenet of achieving good fish capture with good fish welfare. Even when using electric fishing to reduce predator numbers, the welfare of the other, desired fish that will experience the electric field needs to be considered. For broodstock collection, particular care needs to be taken as high stress levels can lead to chemical changes in fish tissues that can affect the viability of gametes. For this reason, unless fishing in very high-conductivity systems, all broodstock collection electric fishing should use direct current (DC) waveforms and large anode diameters.

7.1 Operator skill and fishing and processing methods

A skilled electric fishing team will catch more fish than an unskilled team, and Andrus (2000) found that netter prowess was the cause of the greatest amount of catch variability encountered. In addition, the fish will suffer fewer ill effects from the fishing. Inexperienced operators often fish very methodically but slowly; whilst this may sample slow-moving benthic fish very effectively, the more mobile pelagic fish will often be able to avoid capture. However, fatigue in a team can lead to a lessening in efficiency, and sampling plans should take account of this factor.

Good electric fishing technique also comes with experience, and experienced operators often have their own way of doing things that they find works well for them (and the fish). The configuration of net and anode operation can vary between operators. Provided the equipment is suitably designed and lightweight, some operators find it easier to operate both the anode and a catching net. This both adds efficiency in manpower and can result in efficient fish capture (using the anode to draw fish to the net can be more coordinated when done by one person).

Electricity in Fish Research and Management: Theory and Practice, Second Edition. W.R.C. Beaumont.
© 2016 John Wiley & Sons, Ltd. Published 2016 by John Wiley & Sons, Ltd.

With experience, operators are able to 'read' the river more clearly and thus judge where fish may be found within the reach. Hardin and Connor (1992) found significant differences in the number of fish caught and size composition between different crews manning a boom-boat. They also considered that greater familiarity of the site led to increased yield by a particular crew.

Skilled operators also use techniques that cut down the effort required to gain information on species. Twedt *et al.* (1992) compared results obtained from the selective netting of largemouth bass with the total netting of all fish. No significant differences in indices of population structure (proportional stock density, relative stock density and young-adult ratio) could be found between the two techniques.

When planning the sampling, it is important to determine what it is you wish to sample. If you wish to sample the large mobile species present, then the fishing technique and methods used will be different compared with if you want to get good population data for all species present.

Fishing is often carried out in a noisy environment (flowing water and running generators), so verbal instructions are often difficult to hear or understand. For this reason, a system of hand signals should be agreed upon and used for the basic fishing commands, with whistles carried for more strident or emergency use.

Before actually fishing a site, a full risk assessment should be carried out. This should include factors such as access to the water, the water depth and flow, danger from slippery substrate or sharp rocks and so on. All team members should be briefed as to their roles and the communication signals agreed. Only then should fishing commence. The risk assessment can also be used to document the electrical output settings used.

If fishing equipment is to be bank-based, a level area for the generator should be found. The location should ideally be near fast-flowing water where the cathode should be placed. This ensures that any fish incapacitated by the cathode are carried away. The anode cable should be laid out so that it does not tangle (if it is spread out over a large area, it is less likely to tangle than tight coils). Water conductivity at the site should be either known or measured, and the appropriate anode size and output settings determined and set up. Testing of the equipment should be carried out with all electrodes immersed to avoid the dangerous situation of electrodes being energised and out of the water, which is their most dangerous condition. If the site or conditions are different (e.g. in conductivity) to those previously experienced, an area outside the survey reach should be fished to check that the settings used are both not harming but still catching fish.

On the outer edge of the electric field from the anode, there is a zone where fish can feel the electrical current, but it is not strong enough to create an attraction effect towards the anode. In this zone, the fish are likely to swim away from the anode (the fright zone). To reduce the effect of this, anode operators should not remain constantly 'on' whilst fishing but should fish in a discontinuous manner. This ensures that fish are not constantly driven ahead of the anode. Holding areas should be fished from the edges in, so that large numbers of fish are not suddenly encountered leading to net saturation (where more fish than can be netted are incapacitated). Likewise, in weed beds or undercut banks, if the anode is fished in clear water close to these areas, it will draw the fish out of the areas where netting may be difficult. 'Feathering' the dead-man's switch (i.e. quickly switching the anode on and off) is also a useful method of increasing the drawing effect of the anode.

When netting fish, care needs to be taken not to hit other workers with the net poles. Grievous injury can be caused in this manner, especially if the end of a net pole is pushed into an eye socket. If net operators are inexperienced, then some thought should be given to providing eye protection to those who may be in the close area of the net (bucket carriers etc.). Plastic polarising safety glasses will give some protection and also help to see the fish. Cable holders and bucket carriers should also not crowd the anode and net operators (often a temptation in order to see what is happening) as this will increase the possibility of being hit and limit the movement of the net operators if they need to move downstream.

Netting is a skill but may be quickly learnt. If fishing with minimum output required to just incapacitate the fish, then fast reactions and even greater netting skill are required. In these conditions, the fish are often very fast moving and only briefly incapacitated near the anode. Speed of reaction and good hand–eye coordination are needed for efficient netting. Netting tetanised fish is easier, but fish welfare is likely to be severely compromised and is not desired. The most common cause of missing fish is too slow reactions or netting over the top of the fish due to parallax errors in seeing the fish. Fish should not be held in nets for any length of time and should not be kept in nets whilst netting further fish from the electric field.

Once the fish are caught, they should be transferred into suitably sized buckets or bins. If the equipment is being carried in a boat, then large 'dustbins' of about 80–100 litre capacity can be used. If the gear is bank-based and the buckets are being carried, then smaller (~10–15 litre) buckets can be used. The fish should be regularly transferred into larger bins, particularly if large numbers are being caught. The colour of the buckets (particularly small ones) seems to affect the fish behaviour in them. In light-coloured buckets, the fish seem more active and likely to jump out; in black buckets, they seem much calmer. Whether this is due to them feeling more exposed and thus vulnerable in the lighter ones is uncertain. Care needs to be taken if fish are left in black bins for any length of time in sunlight as water temperature can rapidly rise.

Measuring and weighing fish can be difficult if the fish are active (as they should be, if caught and kept in good holding conditions before processing). Light sedation of the fish can improve fish welfare and operators' stress levels, but regional regulations associated with using even light anaesthesia need considering. Measuring and weighing fish to millimetre and sub-gram precision in field conditions are difficult, time-consuming and often pointless. Unless you are performing detailed growth studies or working with very small fish, measuring

Figure 7.1 Ergonomic (and comfortable) fish processing.

in 5 mm or even 10 mm increments is adequate and faster. Similarly, measuring weights to the nearest 1 or 2 g is usually sufficient (particularly when the weight recorded will vary depending on how much water is on the fish and how hard the wind is blowing on the balance!). If accurate precision of the weight is required, sitting the balance in a box that shields it from the wind can help (Figure 7.1).

Modern electronic measuring boards that can incorporate digital input from weighing balances and even passive integrated transponder (PIT) tag detectors make data recording easier and save manually inputting the data at the office. If the equipment fails, however, information may be lost – a problem that rarely happens with paper-and-pencil records! The robustness of the new technology is improving, though, and this is definitely the way of the future.

If processing large numbers of fish, thought should be given to the comfort of the processing team. Standing for long periods can be very tiring and lead to mistakes or muscle strain. Lightweight foldable tables and chairs can make life far more pleasant and productive.

7.2 Fishing using wading

When fishing shallow (less than thigh-deep) water, it is usual to wade in the stream whilst fishing. As the method involves wading in a zone of electrified water, obviously good waterproof waders are required and good safe working

Figure 7.2 Upstream wading fishing.

practice procedure is instigated. Hands should be kept clear of the water, and if there is a likelihood of hand contact with the electrified water or energised electrodes then 'linesmen's' (electrician's) gloves should be worn. However, safe working practice and appropriate equipment settings should negate the need for these.

Figure 7.2 shows an ideal personnel deployment when fishing. The anode operator is also netting but (as he is holding the anode in his left hand) will concentrate on netting fish to the right of the anode. The second net operator is covering the left of the anode and is also slightly downstream of the anode (where fish are likely to be swept). The bucket and cable carriers are well back out of the danger zone from the net handles, and the fifth person is on the generator and control box and is out of sight. Wading in an upstream direction ensures that incapacitated fish are not swept away ahead of the operators, where they may be subjected to repeated shocks as the operators progress downstream. It also allows any silt disturbed by wading to flow away from the catch area and not hinder visibility of the immobilised fish. Wading also allows speed of fishing to be controlled and therefore also allows sampling of all available habitat types.

In clear water with hard, non-silty substrate, fishing can also be carried out in a downstream direction. This method can be very effective as fish reaction is greater if the anode field impinges on the head of the fish (as opposed to upstream fishing, where the tail of the fish is likely to be the first part to encounter the electric field). If hand or banner nets are placed strategically in runnels of water and the electrode fished down to them, the method can be particularly effective.

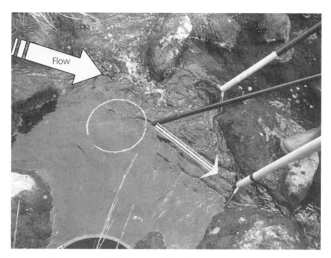

Figure 7.3 Upstream–downstream fishing a small upland pool. Hand nets stop off escape routes, and the anode is fished in a downstream direction (yellow arrow) towards them, thereby both driving fish to the nets and allowing incapacitated fish to drift into the nets.

A variation of this method can also be used whereby even if wading upstream, the stream can be fished in a downstream manner (upstream–downstream fishing). In this circumstance, the downstream exit of small pools or runs can be netted off and the anode fished down to the nets. The team can then walk upstream to the next area and repeat the procedure (Figure 7.3).

Whether upstream, downstream or a combination of the two, sites should be fished in a 'zig-zag' fashion (Figure 7.4a) with the distance between transects approximating the capture field diameter. This will ensure that all habitat types are fished and the whole stream area is covered.

When the stream or river is over 5 m wide, multiple anodes should be used if efficiency of fish capture is to be maintained (Figure 7.4b–c). If the anode safety switch system is of the 'one off, all off' type, then good communication between anode operators is required in order to maintain safety and efficient fishing technique. For very wide rivers, fishing electrodes powered by multiple individual generators may be used, but extreme care and strict safety measures must be taken (see Section 5.1.1).

7.2.1 Wading fishing using boats

When there is dense marginal vegetation or other bank access problems for placing generators or running anode wires, the electric fishing gear can be placed in a shallow-draft boat or punt (often called 'tote boats'). Inflatable boats can be used, but care needs to be taken that the exhaust does not blow on the rubber skin of the boat. In shallow, wadeable water, the gear can then be towed or pushed behind the operators (Figure 7.5). Pulling the boat has the advantage that there is not a person at the rear of the boat, where the cathodes normally

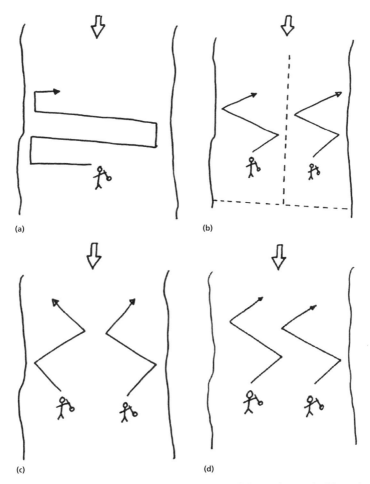

Figure 7.4 Fishing methods when wading. (a) Single-anode fishing. This method is particularly good for population assessment of benthic fish. The distance between horizontal sweeps across river is based upon an effective anode field diameter. (b) Twin-anode fishing, alternative method. The mid-river net lanes off the river, allowing less chance of fish avoiding anodes and escaping downstream. (c) Twin-anode fishing, poor method. When anodes move apart, fish have an easy escape downstream. (d) Twin-anode fishing, good method. When one anode moves to the margin, the second anode covers mid-river, preventing an easy escape downstream.

trail. If the person at the rear of the boat fell in the water, they would be in an electric field and the anode operators may not immediately be aware of their predicament and not release the anode switches. It is important that good access to the emergency stop buttons is still possible whatever method is used and that all operators in the water keep a very vigilant awareness of each other so that equipment is shut down the moment someone gets in difficulty or falls in. The method also has the advantage that teams are not constrained by cable length to fishing fixed lengths of river before having to stop and move everything upstream.

Figure 7.5 Wading fishing using a boat to carry the electric fishing gear. Left: pushing the boat; and right: pulling the boat.

7.3 Fishing from boats

When the water is above thigh depth, it is easier to electric fish while in a boat. Standard hand-held equipment can still be used, and the boat can either be towed using ropes from either bank, rowed or powered by a motor. When fishing rivers from a boat, it is usual to fish in a downstream direction with the boat moving just faster than the water current. In this way, any fish incapacitated ahead of the boat will be overtaken and may be caught. In reality, fishing all but the slowest rivers (especially with engine-powered boats) requires great skill and control by the boat driver, including using rowing or reverse gear to slow and steer the boat.

Care needs to be taken with the position of the cathode so that it does not drift to where the anodes could inadvertently touch it and potentially overload the control box. Or, if using an engine, the cathodes should be positioned such that they do not get entangled in the propeller.

Punts can be used when fishing slow or still streams or canals (Figure 7.6). They are usually controlled by ropes to people on either bank or by oarsmen in the boat. They are not suitable for when there is any appreciable flow due to the problems of controlling the weight of the boat in flowing water. They are normally fished in a zig-zag fashion with the rope operators or oarsman pulling to-and-fro across the stream whilst also controlling the punt's progress along the river. Heavy marginal vegetation will make operation by ropes difficult if the stream margins are too deep to wade.

Small inflatable boats can also be used for fishing where there is some river current. For greater control and manoeuvrability, flat-bottomed boats (without inflatable keels) are preferred. The boat should have decking boards fitted to give a stable platform for the generator and anode operator. The boat is rowed stern-first, so the oarsman can see what is happening and steer the boat accordingly; this takes some skill to perfect. In weedy conditions, the boat will skim over the weed (another advantage of the flat bottom), but the oars need to be kept at a shallow angle in the water so they do not become entangled and impede progress. In rough water conditions or with inexperienced oarsman, standard

Figure 7.6 Left: Fishing from a punt, rope operated; and right: oar operated (right photo with permission Paul McLoone, IFI Ireland).

Figure 7.7 Fishing from an inflatable boat. Note the safe position of generator exhaust and clear access to the stop button of the pulse box (just behind the oarsman). Also note how the small lightweight net is still able to catch large fish (Atlantic salmon).

hand-held equipment can be fished whilst kneeling on the stern of the boat; in calmer conditions, the anode operator can stand in the boat (Figure 7.7).

Where conditions require the use of engines on boats (high flows or long distances to cover between launch and fishing site etc.), then hand-held gear can still be used, but consideration should be given to using boom-boats (see Section 7.3.1).

7.3.1 Boom-boats

In large water bodies, the use of a boom-boat should be considered. These boats have the anodes mounted on booms attached to the boat (hence the name), which are then positioned at the front of the boat where a net operator, also positioned at the front of the boat, can net any incapacitated fish. Anodes usually consist of Wisconsin ring arrays but can also be metal spheres. Cathodes need to be of sufficient size to optimise the anode field, not in an area to which fish are brought into close proximity, and also need to be kept away from outboard engine propellers. If the boat is of metal (usually aluminium) construction, the boat hull itself can be used as the cathode. This gives a good surface area of cathode and thus low electrical resistance. Some countries and agencies prohibit the use of metal-hulled boats due to a perceived greater risk of electric shock to operators (Hickley 1990). Part of the concern is that the boat could become 'live' as a result of a short-circuit in the equipment and electrocute the occupants (Goodchild 1990). In addition, because the metal of the boat is being used as the cathode, it will be at the voltage potential of the cathode (by the way, this will still be the case even if separate cathodes are suspended from the boat, due to electrical coupling of the cathodes and boat hull through the water). If an operator touches the hull with bare skin (or non-electrically insulating clothing), then they too will be at the cathode potential. They will not receive a shock, however, unless they simultaneously touch another object or surface at a different voltage potential. For example, operators may get a shock if they touch the water or a metal out-of-water object that is also earthed (e.g. a metal bridge) and the hull at the same time. In the former case, this is likely to give a shock even with GRP-hulled boats. However, if operators are wearing rubber boots, working in metal boats is no more dangerous than wading in water (that will be at the anode potential) when using conventional anodes. As with any electric fishing equipment, if the equipment is considered 'live' while the generator is running and safe working practices are adopted, there should be no additional safety problems over nonconducting boats.

Cowx *et al.* (1990) compared the efficiency of a multiple-electrode fishing boom and conventional hand-held electric fishing equipment in a series of canals. The boom array was a series of 'dropper' electrodes arranged on a linear boom, and it was fished at right angles to the river flow. The boom array produced better and more consistent catches than the conventional hand-held anodes and was considered a good, cost-effective method for sampling large, linear water bodies.

Boom-boats are very manpower efficient, with the smaller ones (Figure 7.8) only needing two persons to operate (one nets-man and one engine and pulse box controller). Larger boats usually have two nets-men and can carry up to five personnel, to allow for fish processing and the like. Launching the larger boats can be a problem, and purpose-made or commercial slipways may be needed; however, as the larger boats can also carry much larger engines, they can make fast progress between launch and sample sites (Figures 7.9 to 7.11). Jet drives allow fast passage in shallow water without the risk of damaging propellers.

All metal items (e.g. generators and guard rails) in an electric fishing boat (of any construction material) should be electrically bonded to each other to avoid

Figure 7.8 Mini boom-boat using a flat-bottomed punt (allowing a very shallow draft), twin 600 mm Wisconsin ring anodes at the bow and twin 1.5 m cathodes hanging from each side of the middle of the punt.

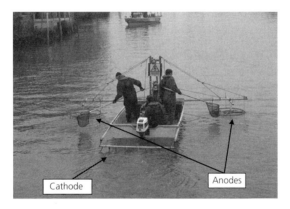

Figure 7.9 Medium-sized boom-boat using a GRP cathedral hulled boat (for good stability), twin 1 m diameter Wisconsin ring anodes and a cathode array mounted on booms beyond the rear of the boat to minimise electrical coupling of the anode and cathode.

the risk of shock when touching two items that may be at different electrical potential to each other. This is particularly important when using boats with metal engines immersed in the water, as the engine will have the same electrical charge as the water.

Separation of the booms (and thus anodes) affects voltage gradient profiles in a similar way to the separation effects when using twin hand-held anodes. Miranda and Kratochvíl (2005) found that the optimal separation of 0.9 m diameter twin anodes was 1.9 m with a suggested maximum of 2.5 m.

There are few studies on the effectiveness of boat electric fishing in non-wadeable rivers. Meador (2005) compared species richness from single-pass and twin-pass 'boat' electric fishing and found that only 65% of species were

Figure 7.10 Medium-sized boom-boat using a metal (aluminium) cathedral hulled boat and twin 1 m diameter umbrella-style Wisconsin ring anodes. The engine size is 25 hp and is suitable for slow-flowing rivers, canals and small lakes.

Figure 7.11 A large boom-boat with light arrays for night fishing and a large 200 hp jet-drive engine (photo with permission Jon Museth, NINA Norway).

sampled by the single fishing. The problem became more pronounced when more than 10 species were present.

Chick *et al.* (1999), in assessing boom-anode-equipped airboats for fishing platforms, considered that for large fish in shallow vegetated habitats, the equipment gave good abundance indices but length-frequency and species-composition data should be treated with caution.

Bowles *et al.* (1990) compared the capture efficiency of a boom-boat (fitted with a 10 m boom with an array of (vertical) tubular electrodes mounted at 1 m intervals) and three punts with personnel operating standard twin anodes. Using marked fish introduced into the fishing reach (which was demarked by stop nets), the three-punt method captured just 11% and the boom-boat 36% of the marked fish. However, both methods gave population estimates well below the number of fish stocked into the fished reach. Welton *et al.* (1990), using the same design of boom-boat, found that capture efficiency based on Zippin maximum-likelihood calculations (Zippin 1958) varied between species (e.g. from 33% for grayling (*Thymallus thymallus*) to 52% for barbel (*Barbus barbus*)), and this efficiency also varied between sites on the river.

Overall, whilst boom-boats are a very manpower-efficient method of fishing larger water bodies, it can be very difficult to achieve good, low confidence limit, fish abundance data. Therefore, in large water bodies, this type of equipment is best used for qualitative or relative abundance data. As part of an 8-year study of a 25 km stretch of the river Tees in Northern England, boom-boat data clearly showed a species shift associated with the construction of a tidal barrage in the lower river (author's data).

CHAPTER 8

Electric fishing 'best' practice

Despite the range of literature detailing negative aspects associated with the capture of fish using electric fishing, it is generally considered that, provided the technique is carried out in an optimal manner, the technique should continue to be used for sampling fish populations. This manual deals with both the theoretical and practical aspects of electric fishing, equipment design and use, fishing techniques and fish welfare implications associated with catching and handling fish.

This chapter, while summarising ways in which optimal settings can be achieved for fishing, does not provide a magic wand for the end of all capture-related injury to fish. The information has been based on 'best practice' as determined from published literature, the experience of a wide range of electric fishing practitioners and the personal experience of the author. If applied intelligently to any particular situation, it will reduce the negative impacts that may be associated with electric fishing. Where the populations are rare or endangered, methods other than electric fishing should be considered.

The guidelines are applicable to all types of electric fishing, from wading the smallest stream to boom-boating the largest, and from low-conductivity to high-conductivity rivers.

In general terms, there are two choices regarding equipment set-up for electric fishing. The equipment can be set up to cause the least possible damage to the fish, or the equipment can be set up to capture the highest proportion or number of fish. Rarely do these two set-ups coincide. Knowledge of the theory behind electric fishing can help bring closer the two options.

This chapter lists the key points, options and techniques that should be used to both maximise capture but also achieve low incidence of fish, and human, injury.

Most evidence suggests that alternating current (AC) does cause more injuries than direct current (DC) or pulsed DC (pDC), and therefore **AC waveforms should not be used for fishing unless warranted by specific circumstances**

Electricity in Fish Research and Management: Theory and Practice, Second Edition. W.R.C. Beaumont.
© 2016 John Wiley & Sons, Ltd. Published 2016 by John Wiley & Sons, Ltd.

(e.g. use of fishing frames, a pre-positioned area sampler (PPAS) or fish to be killed). **Where** *possible,* **therefore, fishing should be carried out using DC or pDC waveforms**.

DC has good anodic galvanotaxis, induces tetanus only in the close vicinity of the electrode and has the lowest recorded rate of injury for any waveform type. Direct current should therefore be used wherever possible. **However, there will be** *many* **situations where it is not possible to use DC** (e.g. high-conductivity water, variable electrical characteristics of stream topography, and poor fish response to the DC field for unspecified causes). **In these situations, pDC fields should be used**. PDC has poorer anodic electrotaxis and tetanises further from the electrode, possibly preventing some fish from reaching the capture zone. **Pulse frequencies should be kept as low as possible**. Snyder (2003) suggests 30–40 Hz or lower but noted that frequencies below 20 Hz may not be good for attracting the fish to the anode. However, there is some evidence that high frequencies may be more efficient for capturing small fry.

All fields should be adjusted to the minimum voltage gradient and current density needed for efficient fish capture. Control box settings should be adjusted to optimise fish recovery. Capture efficiency should be a secondary consideration and can often be offset by carrying out more runs (if depletion fishing). This is an area where some trade-off between fish capture and fishing efficiency may exist. Increasing power (by adjusting the 'power' control on some makes of early pulse box or voltage output in more modern units) when encountering deeper areas of water should be discouraged. Whilst it will increase the capture zone, it will also lead to high gradients near the anode with associated risk to both fish and operators. Increasing pulse width will not increase the field area of the anode, but simply increase the power transfer to the fish within the field and thus lead to higher injury. Many operators use a 'standard' current (amps) when fishing. If this 'standard' has been determined on the basis of past fishing success at those particular sites and lack of fish injury, these standards are probably satisfactory.

Personnel using DC for the first time will need to adjust or modify their fishing technique to account for the much smaller effective field found with DC (Snyder 2003). Calculated field intensity data are good for planning, but on-site, in-water measurements are beneficial to verify the actual intensity and distribution of the electrical field, especially given the importance and potential variation in substrate conductivity. Best practice suggests that settings should initially be carried out based on theoretical considerations and then adjusted based on values actually measured in the stream or river (e.g. by use of a 'penny probe'). Voltage field measurements should be taken using a peak voltage meter or a portable oscilloscope. For square waveforms, a calibrated root mean square (RMS) voltage meter can be used. Part of this set-up process will be the decision regarding what voltage to use. In the past, few pulse boxes in use in the United Kingdom have had this option; however, it is a vital tool for tailoring the voltage gradient to ambient conditions. Voltages can be reduced when having to use small anodes in small, high-conductivity streams or increased in low-conductivity streams. Note that there is no physiological reason for 200 V to be the default voltage as lower voltages will often be equally effective in producing adequate field intensities: the author has successfully used 150 V (at 40 Hz and 30% pulse width) in 450 μS.cm^{-1} (specific) conductivity water using a 400 mm anode ring.

The anode head size should be as large as possible. This guidance is particularly aimed at hand-held anodes but equally applies to boom-mounted ones. In practical use, anode diameters of up to 400 mm can be comfortably operated by users. Cylindrical anodes should not be used due to the high current density at either end of the tube. If using DC, available power may influence the size of anode that can be used, but if using pDC, available power is rarely an issue. For hand-held anodes, the practicalities of handling large anode heads and the physical size of the stream are more likely to be an issue. In small streams, if small physical anode size is required, voltage gradient levels need to be checked to ensure that damaging high gradients are not produced. Adding metal mesh to the anode can reduce the consequential high-voltage gradient that will then exist in the vicinity of the anode. Fishing teams should have a variety of anode sizes so that appropriate-sized ones can be used.

Anodes should not be used for actually capturing the fish as this will hold fish in areas of high-voltage gradient.

The cathode should be as large as possible. Although the commonly used braid design of cathodes is electrically inefficient, it is easy to transport and deploy. To provide a cathode of suitably low resistance, lengths of braid of at least 3 m should be used. When using backpacks, however, this may not be feasible due to problems with towing a trailing cathode. In this case, operators should be aware of the effect an inefficient cathode will have on the anode voltage and catching field, and adjust settings accordingly. If multiple anodes are used, the cathode area will need to be further increased. If this is achieved by increasing the number of cathodes, care needs to be taken to ensure that they are separated from each other. If they are in close proximity, they could be electrically coupled and less efficient. Wisconsin ring style anodes pose a particular problem for getting a big enough anode–cathode area ratio. The use of metal boats wired to act as the cathode is an effective way of helping this problem but may have safety considerations. Knowing the electrode resistance of both anode and cathode will allow an assessment of requirements.

Fishing technique using DC and pDC. When using DC or pDC, fishing should be conducted in a discontinuous fashion. This prevents a continuous 'fright response' gradient on the edge of the field keeping fish out of the capture zone. Discontinuous fishing will also use the element of surprise, improve capture efficiency and thus reduce the likelihood of herding fish into the stop net and thus biasing the capture efficiency of the first catch, so called 'front-loading' (Bourgeois 1995). When in close proximity to areas such as clumps of weed, tree roots or other likely refuges, operators should switch on near to, but not in, the area. Fish will be in the attraction zone, and this will have the effect of pulling the fish out from their refugia to where they can be captured. Care should be taken not to have the anode too close to refugia when switching on as the fish may then be in an immobilisation field and will not be drawn from cover. Sweeping the anode when in areas of open water may encourage fish to seek out areas such as weed beds and so on, where again the above technique can be used. When using twin anodes, this discontinuous method may become difficult if local safety regulations require both anodes to be powered simultaneously. This problem can lead to the dangerous practice of keeping the anode live whilst lifting it from the water: **under no circumstances should this happen**.

Unlike DC, the tetanising zone of pDC extends some way out from the anode. Thus, when using pDC, extra care needs to be taken that the anode is not so close to the fish that the fish is instantly in the tetanising zone of the field or that the fish is tetanised whilst still outside the catching zone. However, this aspect can be minimised by using the optimal anode radius suitable for the conditions being fished.

Actual techniques used will vary between running and still waters. In still waters, the fish are far more likely to be able to escape the voltage field. This can be reduced either by fishing next to the bank (to trap the fish against the bank) or by enclosing sections of still water with nets (similar to the technique used for wide rivers).

Generally, electric fishing teams work in an upstream direction. This reduces the problem associated with stirred-up silt impeding visibility. When fishing from boats or in certain low-silt streams, fishing may be efficiently carried out in a downstream direction, particularly small mountain streams where stopping off runnels between rocks with hand or banner nets and fishing down to them is particularly effective.

When fishing wide sites, multiple anodes can be used. Optimal fishing techniques that can be used are shown in Figure 7.4. Zig-zagging upstream when fishing allows random or target habitat types across the width to be sampled. Moving anodes side to side and up and down when fishing, to 'draw' fish, will also help. When using twin anodes in wide rivers when only part of the width is being covered, it may be advantageous for the mid-river anode to move slightly ahead of the bankside anode. This technique will tend to scare the fish into the bank and make capture by the bankside anode more effective. **In general, one anode for every 5 m of river width has been found to be effective for quantitative electric fishing surveys of whole rivers**. If fishing using multiple power sources, great care needs to be taken with regard to safety procedures in case operators fall into the area being fished.

Fish should be removed from the electrical field as quickly as possible. Holding fish in a net out of water is poor practice and considerably increases the oxygen debt of the fish. Holding fish in water subjects the fish to multiple exposure to the electric field and should be avoided. While length of exposure to the electric field does not appear to increase trauma, length of exposure does increase stress levels. Repeated immersion of fish into an electric field has been shown to increase blood lactate levels and thus will increase post-exposure muscle acidosis. However, although the practice should always be discouraged, most operators acknowledge that at times it is unavoidable.

Regarding the nonelectric considerations when fishing, five major issues arise: water depth, water temperature, water visibility, communication and fish welfare.

Electric fishing by wading is limited to the depth in which wading can be carried out safely. The UK Environment Agency Code of Practice states that thigh-deep overall depth and a hip-deep maximum should be used as the criteria. In deeper areas, boats should be used.

Fishing should not take place in conditions of rain, thunder or lightning. All of these conditions markedly increase the safety risk to operators. Conditions of high wind can also make fishing difficult, due to waves impeding visibility into the water.

Extremes of temperature should be avoided when electric fishing. Most practitioners will avoid the hottest months, but it is also important to avoid the coldest months. However, many fish surveys are carried out in the winter as fish growth slows, allowing better comparison of growth across sites sampled at differing times. In general, there is a trade-off between efficiency (poor at low temperatures) and welfare (poor at high temperatures). Exact effects on fish will vary with acclimatisation of the fish, but in the United Kingdom, a water temperature range of 10–20 °C is preferred for coarse fish and 10–15 °C for salmonid species. If fishing has to be carried out at low temperatures due to logistics (e.g. better between-site growth comparisons), increasing pulse width or voltage gradient may improve efficiency (Lamarque 1967).

The rule regarding visibility required for electric fishing is simple: '**Do not put the anode head deeper than you can see**'. The anode should be visible and ideally close enough to the riverbed for its field to encompass it. The visibility required will vary for different species (e.g. small benthic fish require better visibility than larger mid-water fish). In poor visibility, more runs may be required to achieve adequate population estimates.

Good fish welfare during the holding process requires the utilisation of a variety of techniques. Temperature of holding water is the main criterion determining the steps to take to maximise good welfare. Greater care regarding maintaining oxygen is needed in hot weather. **The use of floating mesh cages is considered to be a particularly effective way of keeping the fish in good condition**. Many practitioners separate eel and other mucous producing fish from the catch. The large quantities of mucous these fish produce lowers the water quality (especially if the fish are held in bins) and 'clog up' other fishes' gills. Holding eel in damp sacking can be an effective and appropriate method to hold them separately from other fish for short periods. Oxygen levels in bins are rarely measured, but supplementary oxygen and/or air is often provided. Oxygen levels in bins can decline rapidly. With an approximately 50% stocking density (40 litres of water for 20 kg fish), oxygen levels can decline to 50% of their starting level in 7 min. This stocking level in bins should therefore be regarded as a maximum and then only for short periods. **Remember that the water needs to be agitated to remove CO_2**. It is possible to supply adequate O_2 with a fine diffuser but still build up toxic levels of CO_2.

Maintaining good bio-security is vital if the spread of non-native, invasive species or pathogens is to be limited. **At the end of fishing, all wet gear should be disinfected**.

Good communication systems need to be in place between anode operators. This is especially true if the 'one-off, all-off' multiple-anode safety system is used or if multiple power sources are used. This system can be plain speech, but in wide or noisy sites, some system of hand signals (difficult if an anode is in one hand and a net in the other), whistles or radio communication is preferable. Modern voice-activated radios fitted to head sets are ideal.

With the recommendations made in this document, it is timely to remind personnel that if equipment settings are altered, then some gear calibration will be required if the new settings are applied to long-term work. Heidinger *et al.* (1983) considered that changes to electric fishing gear should not be made partway through a monitoring programme unless it can be demonstrated that collecting efficiency is not altered. However, if those changes are required for safety or fish welfare reasons, then calibration of the various techniques will allow data comparability and should therefore be made.

CHAPTER 9

Fish population assessment methods

Many books and papers have been written on this subject (e.g. Ricker 1968, Gerking 1978, Bohlin 1982, Cowx 1983, Bohlin 1989, Krebs 1989, Cowx 1990), and this chapter cannot hope to completely cover such a large and complex subject in full. European Standard EN 14011:2003 *Water quality – Sampling of fish with electricity* (Anon. 2003b) also gives advice on standardized European methods and protocols for electric fishing. However, this chapter is a very brief overview of the most common methods of assessing fish population numbers using electric fishing.

Assessing the numbers and/or species diversity of fish within a watercourse is essential for understanding the variation (both natural and perturbed) in fish population dynamics, understanding the production of fish within the system and allowing informed management decisions based on that information (Robson & Regier 1968). The extent to which a quantitative estimate of a population is meaningful, however, depends on the degree to which the population under study conforms to the mathematical model which underlies the estimation procedure (Robson & Spangler 1978).

One of the problems of trying to assess abundance of mixed-species fish communities is the different 'catchability' of the different species and the different size classes of the same species. This different catchability is even more important when using catch depletion methods to estimate the population (Stott & Russell 1979). This is because the depletion method assumes constant catch effort between depletion runs. If this is not the case, then the statistical interpretation of the data may be invalid. Too low or too high capture efficiency can also lead to very wide or overly narrow confidence limits around the estimate, or inaccurate estimates (Bohlin 1982). To some extent, good electric fishing technique can help minimize this problem, and good technique can also reduce the effort required for population estimation by removing the need to carry out additional fishing runs (Kruse *et al.* 1998).

Electricity in Fish Research and Management: Theory and Practice, Second Edition. W.R.C. Beaumont.

The absence of elements or sections of the population due to such factors as spawning or feeding migrations also needs to be considered. This requires that some knowledge of the life history, habits and biology of the fish is available. It may be impossible, for example, to estimate the size of an entire population of fish due to differential catchability of young and old fish, whilst the spawning segment may be readily estimated with considerable precision (Robson & Spangler 1978). Ideally, the data should be stratified into cohorts of equal capture probability (based on experience and judgment). Widely differing size cohorts of the same species of fish should also be stratified and analysed independently (e.g. separate fry and juvenile trout from older fish).

There is some disagreement regarding using electric fishing data from multispecies surveys of low-capture-probability species for population estimation (e.g. Knights *et al.* 2001 vs. Baldwin & Aprahamian 2012). Where in doubt, and if high-accuracy data are required on low-capture-probability species, such species are probably best sampled separately using fishing outputs and methods that increase their capture probability.

Once a unit or sub-unit of the population that it is possible to sample has been identified, Robson and Spangler (1978) recommended a two-stage process:

1 Select a model that requires the fewest assumptions about the behaviour, distribution or sampling.
2 Calculate the minimum sample size required to provide adequate precision of the estimate: this may require a preliminary experiment in order to determine it.

Data on the number of fish present in a stream can take three broad forms: census data, relative abundance data and population estimate data.

A fish census requires that every fish in a reach of stream is captured and an absolute count achieved. For example, if a pond is drained, then an absolute count of the population is possible. To some extent, fish counters provide census data, although there will always be some error associated with the data that does not allow an absolute value to be given. Most studies where census data are given involve poisoning the stream, and rarely is the method attempted using electric fishing. Boccardy and Cooper (1963) were one exception and found that rotenone was 35% more efficient than electric fishing for a census count. As the method is rarely used and likely to be inaccurate when it is, census counts will not be discussed further.

Relative abundance estimates are fish counts that are assumed to be related to actual population numbers, although (with the exception of research papers discussing the method) this assumption is rarely tested.

Population estimates can be obtained when the capture probability of a fish can be calculated. The fish capture probability is used together with the catch data to estimate the actual population size together with a measure of its uncertainty.

9.1 Estimating relative abundance

The simplest form of assessing a fish 'population number' in a watercourse is single-pass electric fishing runs in a section or sections of a river or stream. Provided the data are collected with experienced, trained staff, the method provides a quick method for assessing the presence (or absence) and distribution of different fish species. The data collected will, typically, give species lists for different catchments and habitat types.

If these data are to be used to compare relative abundance between sites or fishing teams, then some sort of standardization is required. This standardization will include gear set-up (including voltage settings at differing water conductivities) and a standardized method of fishing.

European guidelines (CEN regulations, Anon. 2003a) for assessing the Water Quality Framework status of rivers specify standardized reach lengths for assessing fish abundance depending on the size (width) of the river (Table 9.1). These guidelines require only that a single run be carried out (index of relative abundance). Where practical or appropriate (specifically for the smaller water bodies), however, multiple fishing estimates can be carried out. The methods also stipulate that sufficient sites are sampled for valid data on the target species. These sites should also be distributed within the water course to be representative of the habitat distribution. The guidelines suggest that a minimum of 200 fish are sampled and the minimum sampled area is 100 m^2.

Vehanen *et al.* (2012) assessed the likely error of the CEN-recommended single-pass fishing with standard catch depletion sampling. They found that single-pass fishing increased the likelihood of missing rare species and species with low catchability. However, comparison of metrics such as tolerant versus intolerant species showed high agreement, and they concluded that single-pass fishing was suitable for assessing the ecological status of streams. Pusey *et al.* (1998), however, found that single-pass fishing gave underestimates of abundance and of presence or absence. Hughes *et al.* (2002), examining the river length required to estimate fish species richness when boat fishing, concluded that a distance of 85 times the river width resulted in 95% of species being collected. Collection of all species required about 300 steam widths to be sampled. Bateman *et al.* (2005)

Table 9.1 European recommendations for minimum reach size for assessing fish abundance.

River dimension	Minimum length to be sampled
Small stream, width <5 m	20 m, whole stream sampled
Small river, width 5–15 m	50 m, whole stream sampled
Large river or canal, width >15 m	>50 m of river margin on one or both sides of the river
Large shallow water, water depth <70 cm	200 m^2 sampled
Large water bodies (e.g. lakes)	>50 m of littoral zone sampled

found that single-pass fishing captured about 76% of the estimated population of cutthroat trout with about 10% of the effort and that single-pass electrofishing exhibited a sufficient level of precision to be effective in detecting spatial patterns of cutthroat trout abundance. In addition, time saved could be put into increasing the fraction of habitat units sampled and thus improve spatial coverage.

Hartley (1980a), however, was quite clear in his views of single-pass fishing, stating, 'There is no-way in which a fish population can be estimated from a single fishing, however thoroughly this is carried out'!

Crozier and Kennedy (1994) described a novel method for juvenile salmonid monitoring in small streams and rivers by performing timed (usually 5-min) single-pass surveys. The surveys are carried out using 'open site' fishing (Chaput *et al.* 2005) (i.e. without demarking the reach by stop nets). The surveys should be standardised to survey either just shallow (prime habitat) or representative habitat (areas fished in the proportion that they are present in the reach). This latter method is more difficult to carry out (due to having to complete a habitat assessment prior to fishing) but does not prejudge what 'prime' habitat is.

The surveys should also standardise the 5 min to a 5-min 'fishing' (i.e. anode-energised) time. If this is not done, the time taken handling fish (capturing, transferring to bin etc.) will be proportionately more in sections with high fish numbers than in sections with low fish numbers, and the actual fishing time will be less – reducing the estimated abundance (Crisp & Crisp 2006). Many modern pulse boxes (particularly backpack units) can now be set to count down energised time to give a standardised fishing time at all sites. This method has been found to be very effective in terms of relating to 'true' population estimates and manpower efficiency, with Bateman *et al.* (2005) finding that they caught ~75% of the estimated (from depletion fishing) population of cutthroat trout but with only ~8% of the effort. The reduction in manpower allows a greater number of sites to be assessed and also reduces the multiple shock effects on the stream biota.

Point abundance sampling by electrofishing (PASE) can also be considered as a single-pass type of survey, with many researchers (e.g. Nelva *et al.* 1979, Persat & Copp 1990, Garner 1997, Perrow *et al.* 1996) recommending PASE as an efficient and cost-effective method for assessing fish abundance and population structure. Several studies have compared PASE catches with population density data obtained from catch depletion methods, and they concluded that this sampling method provided reproducible and quantitative samples and hence allowed spatial and temporal comparisons within and between reaches (Copp 1989). There is some debate about the number of PASE samples required to provide reliable estimates of population density. Garner (1997) proposed a minimum of 50 samples. However, Laffaille *et al.* (2005) found that 25 samples gave good population estimates for European eel populations in shallow streams.

Sample size is related to whether the fish are heterogeneously or homogeneously distributed (clumped distribution or evenly spread). If the former, then far more samples will be needed to account for spatial variation in catch.

For both timed and PASE sampling, data collection is usually relatively quick (compared with 'true' population assessment), and proponents of these methods consider that a large number of small random samples provide more precise estimates and are more significantly robust than a small number of larger samples.

Relative abundance data can provide a measure of (relative) population change through time, provided that the sampling protocol and effort are consistent over time. If the data from the relative abundance methodology are to be extrapolated to 'real' population estimation, it is important that the methods are calibrated against data on the true population number. This calibration includes adjustment for the efficiency of capture for different fish species and the efficiency of different fishing teams, and thus will be specific for those species and those teams. If, for example, one team is fishing a 400 mm anode and 3000 mm cathode and another a 400 mm anode and 1500 mm cathode, then the numbers of fish that the two teams catch will differ simply due to different capture areas of the anodes.

Whilst 'standard' multivariate statistics will provide calibration data, they are generally simplistic and may not include all the parameters that will affect the calibration. Recently, modelling using Bayesian methods has been shown to be very effective for calibrating relative abundance data to estimates of actual population size (Wyatt 2002). The advantage of the Bayesian approach is that it can account for a wide range of factors that affect the uncertainty of the estimate. However, results can be constrained to remain within expected limits by applying prior assumptions regarding the expected range of the controlling factors. The model provides a probability distribution of the calibration between the two data, called the 'posterior distribution', and thus allows the error of the calibration to be assessed. Dauphin *et al.* (2009) used Bayesian modelling to inter-calibrate the relative abundance estimates from the 5-min samples with population estimates from 5-min removal sampling and found that although actual population estimates obtained from abundance data were relatively imprecise, they did allow differentiation of fish density. Brun *et al.* (2011) also used a Bayesian modelling approach to calculate population estimates that are comparable despite different data collection procedures (e.g. methodology or staffing).

9.2 Estimating actual population size

Methods of assessing the 'true' size of the fish population can be divided into two broad types: capture–mark–recapture and catch depletion methods.

9.2.1 Capture–mark–recapture estimates (CMRs)

This method requires (as the name suggests) that a sample of fish from the population to be assessed is captured, marked and released, and then another sample from the same population captured again.

Two variants of the method exist: the open model that allows for recruitment and mortality (e.g. the Jolly–Seber method; Krebs 1989), and the closed model that assumes no (or negligible) recruitment, emigration or mortality in the time interval between sampling (e.g. the Lincoln or Petersen methods). The former is often used for estimating survival rates, often at the catchment scale, and will not be discussed further. The latter is the most commonly used for population estimation in standard electric fishing reaches (~1000 m²).

The closed method assumes that the fish come from a discrete reach in which there should be no immigration, emigration or mortality. For small reaches, this will require that the reach be enclosed by well-set stop nets (due to edge effects in small reaches having a proportionally larger effect than in longer ones). The captured fish are then marked so that they can be identified. This mark can be a batch-mark (fin-clip, dye mark, uncoded tag etc.) or an individual mark (e.g. PIT tag), the important criteria being that there should be insignificant mark/tag loss and the fish are identifiable. They are then released into the netted-off section of stream, and a period of time is allowed in order that they can redistribute themselves within the reach. The reach is then resampled, and the proportion of marked fish ascertained. The second sampling can use a different sampling method or different electric fishing operators, and can have a different capture probability to the first fishing, although capture probability within a single fishing should be the same. Carrier *et al.* (2009) proposed a method where the first sample is taken using traps and the second sample by electric fishing. This reduces the bias in capture probability that may occur if both samples were by electric fishing, due to fish that avoid capture becoming less vulnerable to subsequent capture, (Peterson & Cederholm 1984). Bohlin (1989) considered that using different methods reduces the size selectivity of the capture methodology, with one of the CMR assumptions being that the marked and unmarked fish have the same probability of capture. The method also assumes that the marked fish are normally distributed into the population (i.e. they have remixed) at the time of the second sampling. This latter assumption is obviously a time and sample reach size-dependent function, and ensuring its validity is difficult.

The estimated population number can be calculated from:

$$N = (M+1) \times (C+1) / (R+1) \tag{9.1}$$

where N is the population estimate; M is the number of marked fish in the first sample; C is the total number of fish captured in the second sample; and R is the number of marked fish captured in the second sample.

It is important that accuracy (i.e. the closeness of the calculated value to its true value) and the level of precision (i.e. how much error around the calculated value is acceptable) for the population estimate are specified prior to sampling in order to ensure an adequate sample size. The method produces most accurate estimates when a reasonable percentage of the population is marked and a good precision when a reasonable number of the marked fish (ideally above 10%) are recaptured (Robson & Regier 1964, Krebs 1989). The method is often considered to be independent of sampler and environmentally induced biases (Peterson & Cederholm 1984). Practical problems occur when the method is used in reaches isolated by stop nets. Keeping the stop nets free of weed and debris over the time it takes to allow the fish to redistribute themselves can be difficult, particularly in high-gradient or high-weed-growth streams. The other problem is determining the time needed for the fish to redistribute themselves. Depending on the size of the area sampled, the nature of the stream and how they are released into the stream (evenly distributed or single-point released), this could take from a few hours to several days.

By the nature of the method (requiring at least two fishing runs with significant time delay between each survey), it is not suitable for rapid one-off surveys.

9.2.2 Catch depletion estimates

This method, also known as quantitative or removal depletion, requires that a closed section of stream is fished and all the captured fish counted and removed into holding tanks. The stream is allowed a period of time to settle (depending on reach size from 20 min to 1–2 h, with Bohlin (1989) recommending 'at least half an hour'), then fished again and the number of fish again recorded.

Different variations on the method use either a linear regression method (e.g. Leslie & Davis 1939) to extrapolate the line formed by the successively reducing catch or a more complex calculation using *maximum likelihood models* (MLMs) (e.g. De Lury 1947, Zippin 1958, Seber and Le Cren 1967) or *maximum weighted likelihood* (MWL) (Carle & Strub 1978) that is based on capture probability. Of the three, the linear regression is the simplest and the MWL the most robust (particularly for data where capture probability is variable).

The minimum number of electric fishing 'runs' required to get the MLM depletion estimates can be just two (Seber and Le Cren 1967, Robson & Regier 1968), although three or more runs are often used where the depletion or drop-off in successive numbers capture is relatively small. The calculation of population number from the two-run method is also very simple (Seber & Le Cren, 1967):

$$N = c_1^2 / c_1 - c_2 \qquad (9.2)$$

Or there is the Robson and Regier (1968) model, which reduces statistical bias (Cowx 1983):

$$N = c_1^2 - c_2 / c_1 - c_2 \qquad (9.3)$$

Other MLM and MWL models require more runs and far more complex computations to get the population estimates (Zippin 1958, Carle and Strub 1978, Ricker 1975). Poor catch depletion patterns may require five or more fishing runs if sensible confidence limits to the population estimate are to be attained.

Catch depletion has the benefit over CMRs that the time taken to complete a survey is drastically reduced as a smaller time period can elapse between fishing. However, the method has several assumptions that must be adhered to for the estimate to be valid.

1 The capture probability of individual fish should remain constant over the different sampling runs (i.e. capturing 70% of the fish present in run 1, followed by only 10% of the remaining fish in run 2 will give invalid results and/or poor estimates).

2 Inherent in assumption (1) is that the effort for each fishing run must be constant. Achieving this can be problematic due to the fish handling time being greater in the early runs, due to the (hopefully!) greater number of fish being caught in the early runs (Crisp & Crisp 2006). As the time needed to fish the reach will not be known prior to the actual first fishing, compensating for this bias may be difficult.

3 Emigration and immigration from the reach, and recruitment or mortality of fish within the reach, should not occur. The reach should therefore be demarked with stop nets (or reaches chosen to make escape from the reach difficult), and time scales between fishing runs should be such that no significant mortality will occur.

4 A significant proportion of the population should be removed during the sampling period. This has often been interpreted as 'catch all the fish on the first run', but this is not the case. The aim of the sampling should be to remove a fixed proportion of the fish (preferably above 60%) in each run. This will give a decreasing trend in numbers caught but still give a high overall percentage (e.g. a 60% catch rate over three runs will capture 98% of the population). This pattern of data is also more suitable for analysis. If numbers permit, the fish can be stratified into size cohorts. This is often useful when small young-of-the-year fish are caught, as capture probability is likely to be lower for these fish. If capture rate is too low (<30%), then more runs may be required to get adequate population estimates with good confidence limits.

In addition to these caveats, variation of fish reaction between successive fishing runs will also affect capture likelihood; fish becoming increasingly resistant to, or more able to evade, capture with successive runs (Cross & Stott 1975, Hartley 1980a, Peterson & Cederholm 1984, Mesa & Schreck 1989). This results in many removal-based estimates of fish abundance being biased (Peterson *et al.* 2004). It follows that successive electric fishing runs, which are equal in every controllable parameter, are in fact not entirely equal in effect. To minimise this bias caused by unequal capture efficiency, a variation of the catch depletion model (MWL model) can be used (Carle & Strub 1978, Krebs 1989). It is also possible to perform statistical tests (often chi-square) to test for the assumption of equal probability of capture between successive runs.

When using catch depletion estimates, it is important to realize that efficiency of capture will vary both between sites and between times at the same site. The fact that a given electric fishing configuration has fished at 70% efficiency in a particular site does not guarantee that it will perform the same way again at the same site a week later or in a different site later on (Hartley 1980a). However, with skilled operators and standardized equipment and methods, this variation is likely to be small.

Whichever method is used, many studies on the validity of catch depletion estimates often find that when only two or three runs are carried out, the population number is underestimated (e.g. Riley & Fausch 1992, Rosenberger & Dunham 2005, Meyer & High 2011). Meyer and High (2011) found that capture efficiency decreased with successive fishing runs, declining from 58% in run 1 to 37%, 30% and 18% with successive runs. For this reason, it is important to accept that population estimates are 'estimates' and not absolutes.

When sampling for population estimation, it is important to get a sample of fish large enough to fulfil the requirements of all the statistical extrapolations performed by the population estimation calculations (Seber & Le Cren 1967). Bohlin (1982) examined the population sizes needed for accurate estimation of confidence limits for the population estimate. Calculations were carried out for both two-sample (using the Seber & Le Cren method) and three-sample methods (using the Junge & Libosvarsky method) at differing probability-of-capture rates. For the two-sample method, population sizes of >200 are needed unless the probability of capture is above 80% where a 'population size down to 100 may be tolerated'. For the three-sample method and capture probabilities between 50% and 70%, the standard error was (reasonably) well estimated for populations down to 50 individuals. However, if the capture probability was >80% or greater, then the confidence limit was smaller than it should be. Bohlin (1982) attributed this to Zippin's (1958) findings regarding skewed patterns of fish capture when high catch rates are encountered.

Fish surveys, however, often involve far fewer than 50 fish, particularly if the catch is subdivided by species and age cohort or when carried out in small streams. In these instances, Bohlin (1982) recommends using pooled estimate of p from several fishings. In these cases, whilst the various population calculations can be carried out, they should be treated with even greater caution: Krebs (1989) commented that 'you should never confuse statistical significance with biological significance', and 'garbage in, garbage out' should be remembered. When numbers are very low (<30 individuals), and particularly when the last fishing run has not caught any fish, then totalling the catch from all runs and describing the population size as, for example, 10 fish/m² 'based on actual catch' probably comprise the most accurate way of describing the population size.

CHAPTER 10

Fish barriers

The use of electricity as a barrier to fish movement can be divided into two broad categories: barriers to stop or divert upstream movement and barriers to stop or divert downstream movement. In addition, some limited use has been made of electric barriers to herd fish into areas where they can be caught by conventional nets (Liu Qi-Wen 1990). Although not exclusively so, upstream barriers tend to block or stop upstream moving or migrating fish entering blind culverts (such as turbine tail races), but they can also be used to prevent upstream access into areas of water where they are not wanted, such as preventing the spread of invasive fish species (O'Farrell *et al.* 2014). Downstream barriers are used mainly for diverting or guiding downstream migrating fish around danger areas (such as hydropower or water abstraction intakes). Of the two types, upstream barriers tend to be the more straightforward to design and downstream guiding screens the more complex; however, hybrid systems can exist.

The screens rely on the avoidance behaviour of fish to a weak, low-current-density electric field (Stewart 1990). This avoidance is the fright response commonly seen on the edge of the (high-current-density) electric field used when using electricity to capture fish. The fish's reaction depends on the electrical potential between its head and tail (the body voltage). This potential is at its maximum when the fish is perpendicular to the field and at a minimum when the fish is parallel to the field (Vibert 1967a,b). Screens using DC or pulsed DC (pDC) fields also use the physiological effect of the fish's inability to face a negative field (cathodic repulsion) to exclude or guide them. In most designs using this effect, the active electrodes in the water are negatively charged (cathodes) and the positive (anodes) can be embedded in the bank of the river or positioned in some other location away from the barrier or diversion screen. However, systems whereby electrodes are alternatively switched between acting as anodes and cathodes are also used (Mishelovich & Aslanov 1990). This switching can both prevent fish from becoming acclimatised to a current density pattern and also reduce electrolytic corrosion of the electrode material.

Electricity in Fish Research and Management: Theory and Practice, Second Edition. W.R.C. Beaumont.
© 2016 John Wiley & Sons, Ltd. Published 2016 by John Wiley & Sons, Ltd.

Electric barriers have the advantage over physical screening in that (provided they have been designed appropriately) they should not be affected by debris, turbidity or high flows. However, the screen's effect is increased when it is used to reinforce some other influence on behaviour (e.g. water flow or the visual stimulus of the electrode array or artificial lights) (Stewart 1990, Mishelovich & Aslanov 1990). Hadderingh and Jansen's (1990) findings of poor reliability of deflection of the electric screen they were trailing at times of high algae bloom and at night may have been due to these factors reducing the visual stimulus aspect of the screens.

Barrier size can range from small 1 m culverts using only a few tens of watts (e.g. Johnson *et al.* 1990) to huge barriers that can be in excess of 300 m long (Zhong 1990) and require kilowatts of power to energise.

Electrode design is of two main types: vertical rods, either fixed into the bed of the water body or suspended from an overhead wire that enables them to move out of the way of drifting debris, and horizontal electrodes mounted on the base of the river or culvert. The fixed vertical electrodes will need a mechanism to remove debris, and the bottom-mounted ones will need to kept clear of accumulated sediment in order not to affect the barrier's efficiency or, if the sediment is very conductive, the power demand of the barrier. Conductive sediment or steel-reinforced concrete will also affect the voltage gradient pattern of both types of vertical electrodes (Lethlean 1953).

One of the first published uses for electricity in fisheries management was a German patent by Larsen in 1910 for an electric barrier (W.G. Hartley, personal communication). The idea was not developed further, however, until H.T. Burkey took out the first of a series of patents in 1917 in the United States. The screens he installed were usually (but not exclusively) an array of vertical rods set across the river as one electrode, with another electrode array fixed on the bottom of the stream parallel to it. They were powered by either AC or DC (supplied from a magneto). The patents, however, did not describe the underlying principle behind how the system worked, so the screens described could not be easily replicated by other workers. In 1929, MacMillan published a paper describing the theory behind the effect and giving practical details of electrode construction and screen design. His favoured design was a double row of vertical rod electrodes of large diameter (to limit excessive gradients close to the electrodes) mounted in parallel with each other and powered by 60 Hz AC. After this, electric screens became more commonly used and investigated. Work on electric screens was also being carried out in Japan (possibly based on a Japanese patent by Takahashi in 1895), but, due to difficulties in access and interpretation, the work was largely unknown to Western researchers.

Historically, the earliest electric screens used AC current, and many are still powered using this waveform. More recently, pDC (both exponential pulse and square wave) has been used, both because of its lower power demand and due to its less damaging effect on the fish. McLain (1957) found that fish injury was reduced from 86 to 8% as a result of switching from AC to a square-wave, 3 Hz, 66%-duty-cycle pDC current. Frequencies used in pDC screens are typically low, and Larson *et al.* (2014) found that increasing screen frequency from 11 to 20 Hz doubled the injury rates observed in cutthroat trout.

Ideally, the barriers should have a graduated current density that enables some control over the severity of the fish's reaction to the field. Fish should have time to develop a defensive reaction and move away from the stimulus (Vibert 1967a,b). A graduated field will also allow the screen to be used for a range of fish sizes with individuals reaching their own, limiting, gradient intensity as they progress through the screen. Ideally, the field current density should also progress from zero to a density that is capable of stopping a fish half the length of the smallest fish: large fish turn away earlier, small fish later, but all should eventually be affected and repelled. Graduation can be achieved either by energising the electrodes at different potentials (Figure 10.1) or altering the size of the electrodes or by adjusting the spacing of electrodes at the same potential (Figure 10.2).

Blocking screens are designed to prevent fish from moving upstream into areas of potential danger or into areas where they cannot continue upstream (turbine tail races etc.). By keeping them away from these dangers, the fish are encouraged to find alternative routes (that should be either present naturally or

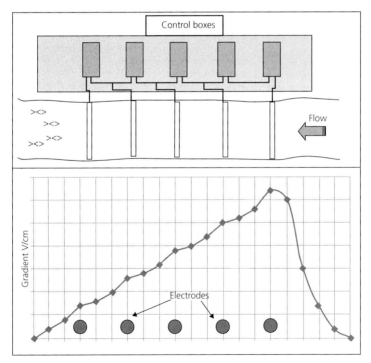

Figure 10.1 An example of how to create a graduated electric field in an upstream barrier by increasing applied voltage to fixed space electrodes.

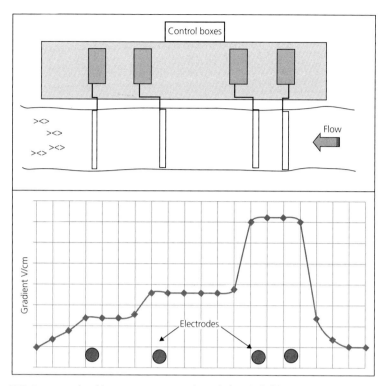

Figure 10.2 An example of how to create a graduated electric field in an upstream barrier by altering the spacing of fixed voltage electrodes for an upstream barrier screen.

provided) around the dangers. The screens can also be used to stop the spread, or contain, invasive or non-native fish species, for example the barrier in the Chicago Ship Canal that is preventing the spread of Asian carp into the Great Lakes of North America. At its most basic design, this type of barrier is relatively simple to design and install. Large installations, however, may have to be able to operate when metal boats go through the barrier whilst still maintaining their barrier efficiency and not affecting the boat. The advantage of an upstream barrier (particularly one using a cathodic field) is that it does not purely rely on just the fish's behaviour in avoiding the electric field, as the fish is physiologically incapable of ascending upstream through such a barrier. Barrier location should have a reasonable flow of water, Burrows (1957) and Johnson et al. (1990) recommending a minimum of 1 meter per second (m.s) or above one body length per second. In such flows most fish will face upstream into the flow, and if they are moving upstream then they must face upstream. If they then encounter an increasing electrical current gradient (negative or positive) with the current parallel to the stream flow, they will be incapable of continuing. As soon as the fish turns across the water flow, the current density experienced by the fish will decrease and the water flow will sweep the fish downstream away from the obstruction. If the fish persist in attempting to move upstream, then the increasing current density will eventually induce incapacitation or narcosis, whereupon the fish will also be swept downstream. If a cathodic current is used, the effect is further enhanced by the inability of the fish to face such a current and its repulsion effect. Due to the predicted orientation of the fish when swimming to be at right angles to the current, a very simple electrode arrangement can be used (Figures 10.1 and 10.2). Current gradient in the upstream barrier can be quite abrupt as there should be no problem with fish being swept downstream in the water flow. If some guidance into alternative upstream routes is needed, then some low-level graduation and strategic positioning of some electrodes can achieve this. Ideally, however, there should be a sufficiently good attraction flow to the alternative route that this is not needed.

Diversionary barriers for downstream moving fish are far more complex to design. They are intended to divert downstream migrating fish away from danger zones such as water abstraction points and turbine intakes. Fish need to experience discomfort from the field whilst still being able to react and turn away from the source of the discomfort. However, if the fish become incapacitated due to the action of the screen, then they will be swept into the area from which they are meant to be excluded. For this reason, it is important that the screens are positioned in relatively slow-flowing water, Zhong (1990) recommending <0.5–0.7 m.s, and O'Farrell *et al.* (2014) 0.3 m.s and for the screen to have a low current-gradient graduated barrier that gives the fish time to react and turn away. Ideally, the electrical current should also encourage the fish to move towards the bypass channels or other escape route. If a pDC gradient is used and a very diffuse current-gradient anode is placed upstream of the less diffuse current-gradient cathode, then an increased stimulus to turn away towards the anode will occur. With diversionary screens where there is a preferred route for the fish to take, the anode can be offset to encourage the fish to turn towards this direction (Figure 10.3).

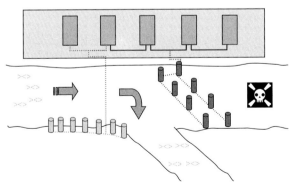

Figure 10.3 A barrier for downstream migrating fish. Blue cylinders are cathodes that are powered sequentially to provide an oscillating electrical field, red cylinders are anodes placed along the bank to help attract fish to the by-wash canal, and green arrows are fish movement.

Downstream screens also have the problem of the fish being able to approach the screen in many different orientations, depending upon whether it is actively swimming downstream or merely drifting passively with the water flow. For this reason, a simple parallel current screen may allow a fish that is drifting downstream sideways (parallel to the electrical current) to enter a relatively high current gradient before being startled, changing orientation and then receiving an incapacitating shock along its head-to-tail orientation, which results in the fish being swept into the danger area. A patent by Baker in 1936 recognised this problem (W.G. Hartley, personal communication), but the published work of Lethlean (1953) described how to use an oscillating electrical field where the lines of current were not fixed but constantly varied in their angle to the flow of water and prevented a fish from achieving an attitude in which it was unaffected by the field. Electrodes can also be energised sequentially to create a moving-vector electric field; this will also have the effect of reducing the overall power demand of the screen as not all electrodes will be energised at any one moment in time.

Because the electric screens are designed to be constantly energised but are unsupervised, there are concerns regarding their safety. Whilst fish mortality has been known to occur, particularly under fault conditions (Johnson *et al.* 1990), it is the effect of the electric screen on humans who may find themselves (either voluntarily or involuntarily) in the electrical field that is of most concern. Even the slightest risk of human injury may make areas of high public access not suitable for the use of electric screens (particularly where an incapacitated person may be swept into an area of danger). O'Farrell *et al.* (2014) states that the maximum pulse width from the screen they were describing was 10 m.s (0.01 s), and this pulse duration is well below the electrocution threshold for a typical adult. Johnson *et al.* (1990) attempted to define the size of the danger zone around an electric screen they had installed in southern England but found that the literature was not appropriate for assessing diffuse source electrical current; concentrating on point source which has more severe effects. They recommended that, using the point source data with a safety factor of two, a 3 m exclusion zone would be sufficient.

CHAPTER 11

Fish counters

Resistivity fish counters, whilst not actively using electric fields to create an effect upon fish (capture or repelling), use electrical principles to detect fish. Fish are detected by passively monitoring the change in electrical resistance in a (confined) body of water when a fish swims over an array of electrodes. Fish behaviour is not – or should not be – affected. The counters were first developed to assess the effectiveness of fish passes around hydroelectric dams but have subsequently been used for monitoring fish (predominantly salmon) populations in a wide variety of locations. Hellawell (1973), Hellawell *et al.* (1974) and Bussell (1978) give an overview and technical details of counter development and operational set-up, and Beaumont *et al.* (2007) gives a more recent perspective.

Lethlean (1953) developed the first commercially available fish counter for use on the North of Scotland Hydro-Electric Board's (NSHEB) dam on the river Tummel at Pitlochry in Scotland. The design had an extensive list of operational requirements, including: no false readings from erratic fish passage or debris passing over the electrodes, the equipment should work with minimal maintenance and the equipment's operation should not be affected (unduly) by seasonal variations in temperature and chemical variations in the water.

The first NSHEB counters used a 'Wheatstone bridge' to monitor the resistance between three circular electrodes placed inside a ~75 cm diameter pipe in a fish pass. The electrodes were configured in two pairs with the centre electrode common to both pairs. The resistance between the electrodes and the Wheatstone bridge is finely balanced using a potentiometer, and in this state no voltage passes through the detecting circuit. When an object of different electrical resistance to the water passes through the electrode array, the resistance becomes unbalanced, and a small voltage in the detecting circuit is created and a 'count' recorded. The voltage thus created is very low; Lethlean (1953) used voltage gradients that were about 30% (approximately 0.02 V.cm^{-1}) of the values that were seen to affect a 40 cm salmonid in the low-conductivity water that he was dealing with. Logic circuitry within the counter unit detects the direction the fish is travelling (based on the time sequence in which the upstream and downstream pairs of

Electricity in Fish Research and Management: Theory and Practice, Second Edition. W.R.C. Beaumont.
© 2016 John Wiley & Sons, Ltd. Published 2016 by John Wiley & Sons, Ltd.

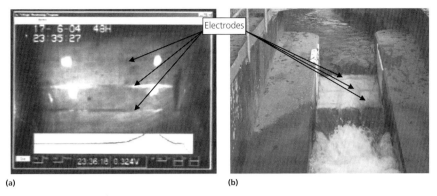

(a)   (b)

Figure 11.1 (a) Counting electrodes installed in a gauging flume showing a fish ascending the weir and part of a waveform trace used for verification purposes. (b) Counting electrodes installed into a fish pass.

electrodes are energised). Electrodes are normally made from stainless steel; however, some steel can have a skin effect that affects the electrical contact with the water at the low-voltage and resistance values being used in counting systems. To overcome this, platinum-coated titanium electrodes have also been used on small smolt counting systems (author's data).

The size of fish that are detected is based on the electrode separation, with a 'rule of thumb' that the electrode separation should be around the minimum length of fish that operators wish to detect. For adult Atlantic salmon (*Salmo salar*), one of the most common species that are counted, an electrode separation of 450 mm is commonly used, giving good detection of a ~500 mm fish that is around the minimum size of a 1-sea-winter fish (Bussell 1978). Some size discrimination (to exclude detection of small fish) can also be achieved by altering the sensitivity settings on the equipment that set the resistance thresholds that will register a count.

Subsequent to Lethlean's counter, NSHEB and other manufacturers have produced counters that work in open channel flumes (Figure 11.1). Water velocity in the flumes should be such that the fish make a clean passage across the electrodes and do not repeatedly partially ascend and descend over the counting array (vacillate). The flumes should also not be subject to wave action that creates resistance changes that mimic the patterns of resistance change created by fish passage over the electrodes. The newer units also have circuitry to automatically adjust settings to cope with gradual changes in water conductivity, which would affect the size criteria thresholds. Modern electronics and microprocessors have also replaced the Wheatstone bridge and often measure current change from electrodes energised at a fixed voltage. Where other equipment is installed in the river that may be sensitive to the electrical field emanating from the counting electrodes (e.g. other counting channels or a passive integrated transponder (PIT) tag detecting equipment), additional guard electrodes (which are referenced to earth potential) can be used upstream and downstream of the counting array to constrain the field.

Figure 11.2 A multichannel resistivity counter used for detecting downstream moving salmon smolts.

The width of river that can be covered by a counting system is dependent on the conductivity of the water; in low conductivity, the relative change in electrical resistance caused by a fish is large, and so it is easily detected and thus the electrode length can be longer. Hellawell (1973) recommended that in low-conductivity systems 10 m, and in high-conductivity ones 3 m (both conductivity values unspecified), of river can be covered by a single electrode array. Several arrays can be installed, however, either on the same level or on a split-level weir configuration (Bussell 1978). In general, the wider the single array, the less the sensitivity of the array and the more false or undercounted counts are likely from multiple fish passage.

Any resistive object of sufficient size (as determined by electrode spacing and threshold resistance limits) will be detected by resistivity counters. It is common for 'fish records' to include weed mats and various dead animals drifting downstream, otters moving over the counting zone as well as non-target species of fish. For this reason, verification of the 'counts' is vital. Historically, this task was carried out by a constant watch of observers (Lethlean 1953), but this was often impractical as well as costly. Humans were replaced by motorised film camera (Hellawell *et al.* 1974), but time-lapse CCTV is probably the most common verification now in use. CCTV only works in clear water conditions, however, so other methods of verification that will work in turbid water (often the time when many fish migrate) can also be used, such as waveform analysis (Bussell 1978, Beaumont *et al.* 1986) or acoustic imaging.

Resistivity counters can also be used to count small fish (e.g. migrating salmon smolts). Electrode separation is reduced commensurate with the smaller fish length – approximately 75 mm is sufficient for Atlantic salmon smolts. The volume of water monitored also needs to be much smaller in order to detect the very small resistance changes that the fish produce – a square 'tube' of 200 × 200 mm is about the maximum size that can be used with present systems. If used in an array such as that shown in Figure 11.2, then addition guard electrodes are recommended to limit the propagation of fields between tubes. These systems have greater problems with false records, however, and so stringent verification of records is needed.

CHAPTER 12

Electroanaesthesia

The incapacitating effect of an electric field can be used to deliberately keep fish immobile for long enough to perform management tasks such as stripping, tagging or measuring. Whilst chemicals can be used for this purpose, there are concerns about both the time taken to excrete the chemicals and the effects of the chemicals and their residues that may enter the human, and to a lesser extent the animal, food chain. The most common chemical fish anaesthetic in the United Kingdom, United States and Canada, and the only one that is allowed to be used in the United Kingdom without an ASPA (Animals (Scientific Procedures) Act of 1986) licence (see Section 13), is MS222. As well as a significant list of possible side effects to the fish, it takes 24 hours to be excreted (via the fish's urine), and the US Food and Drug Administration (FDA) has a recommended withdrawal time (the time delay before the fish should enter the food chain) of 21 days (Ross & Ross 1999). This means that if you use the chemical on a fish that could be eaten, you will need to hold those fish for 21 days before release.

In addition to these issues, chemical anaesthesia (particularly the more common inhalation anaesthetics) can take a significant time to immobilise the fish. This is especially true for air-breathing fish such as catfish and eels.

Electroanaesthesia has the advantage over chemical methods in that it is immediate (or at least very short in its induction time) and allows quick recovery with no withdrawal period required. Sattari *et al.* (2009) found induction-to-recovery times with an AC electroanaesthesia system were about 10% compared with chemical methods (MS222 and clove oil). Gosset and Rives (2005) found that direct current (DC) electroanaesthesia was superior to chemical anaesthesia using clove oil, being immediate and causing less post-operation stress.

When using electroanaesthesia, waveforms and voltage gradients should be such that immobilisation is a result of narcosis (not tetanus) and the severity of the procedure should not result in any physical damage to the fish. As with electric fishing, the gradients needed to immobilise the fish will vary with the conductivity of the water, and different species or sizes of fish will also require

Electricity in Fish Research and Management: Theory and Practice, Second Edition. W.R.C. Beaumont.
© 2016 John Wiley & Sons, Ltd. Published 2016 by John Wiley & Sons, Ltd.

different settings. Waveforms used include AC, DC and pDC; however, modern units tend to just use DC and/or pDC. If just DC is used, the fish are likely to recover as soon as the applied voltage is switched off. The use of a combination of both DC and pDC allows the gentler DC to be used for the initial immobilisation, and then the pDC to be used to provide an increased effect sufficient to keep the fish immobilised after they are removed from the treatment tank. Vandergoot *et al.* (2011) compared the effect of DC and pDC waveform electroanaesthesia and recommended pDC on the grounds that it gave immediate effect with quick recovery time and high survival. Precise control of the duration and intensity of the output is achieved by programming the settings desired in the control unit.

Figure 12.1 shows a small portable electroanaesthesia system (PES) produced by Smith-Root Inc., United States. Waveform type and duration can be programmed for a primary (initial narcosis) and secondary (immobilisation) waveform. Larger units can also be constructed, and the big Pacific salmon hatcheries frequently use systems designed to immobilise several large adult salmon simultaneously.

Most small units use such low-voltage gradients and power densities that the fish can be handled with bare hands in the tank – although gloves are recommended.

A recent development in electroanaesthesia is the DC Fish Handling Gloves developed by Smith-Root. These are gloves that are energised with an appropriate DC voltage, so just handling the fish will induce narcosis in the fish.

There are some concerns that the method does not produce neurogenic effects in the fish (i.e. the fish can still feel pain while immobilised). Also, there

Figure 12.1 Small portable electroanaesthesia unit made by Smith-Root Inc., United States.

are mixed reviews of the effectiveness and settings required for immobilisation, with some researchers finding significant haemorrhaging in certain species.

Marking and Meyer (2011), when reviewing researchers' thoughts about what made an ideal anaesthetic agent, concluded that the ideal anaesthetic should allow a reasonable time with the fish immobile, have an induction time of 3 minutes or less and a recovery time of 5 minutes or less, cause no toxicity to fish or mammals and have a withdrawal period of 1 hour or less. On that basis, electroanaesthesia would seem to be an ideal candidate. However, further research and guidance on suitable settings are still required.

CHAPTER 13

Fish welfare

Proper handling of the fish once caught is essential, both to prevent death or injury and to reduce stress. In the past, considerations about a fish's ability to 'suffer' have been somewhat overlooked. However, recent research has shown clearly that fish can react to stressing actions, and some research surmises that fish can not only feel pain but also experience fear (Verheijen & Flight 1992). Other researchers dispute that fish can feel pain or fear (the fish not having the brain structures required to recognise these factors) but do acknowledge that they display 'robust, nonconscious … stress responses to noxious stimuli' (Rose 2002). Although the debate continues regarding this issue, fishery workers must be aware of the fact that they are dealing with sentient organisms and act appropriately. If killing fish is required, then it should be done in a humane manner (e.g. anaesthetic overdose and/or cerebral maceration). Fishery workers should be aware that many countries and organizations have guidelines regarding what is ethical to do to animals in the name of research or management. In many cases, this will just be an ethical review board based in the organization or university or other competent authority. In the United Kingdom, such work is covered by the Home Office under the Animals (Scientific Procedures) Act 1986 (ASPA). The UK ASPA legislation states that when fish are being used for a scientific purpose, and if this may cause pain, suffering, distress or lasting harm, project and personal licences from the Government (Home Office) are required. In Europe the new EU directive, which updated the UK ASPA legislation on 1 January 2013, states that fish in scientific studies are protected under law from the point at which they become capable of independent feeding.

Where procedures are regulated, acquiring the correct licences to perform procedures under those regulations often involves specific training. There are also a number of other requirements that need to be met, and therefore there may be issues of practicality for some workers. In the United Kingdom, if you are in any doubt as to whether the work you are performing might need licences under ASPA, you should contact the Animals (Scientific Procedures) Inspectorate Aquatics Group at aspa.dundee@homeoffice.gsi.gov.uk.

Electricity in Fish Research and Management: Theory and Practice, Second Edition. W.R.C. Beaumont.
© 2016 John Wiley & Sons, Ltd. Published 2016 by John Wiley & Sons, Ltd.

13.1 Fish handling

Handling wet, slippery, wriggling fish is a skill. When handling fish, hands should be cool and wet and all handling kept to a minimum. Fish are coated with a protective layer of mucous that protects them from infection, certain parasites and the osmotic effects of the water. Experienced handlers may catch fish from holding buckets with their hands; however, care must be taken not to damage the skin (including the mucous layer) as this leaves the fish open to infection and disease. Fish should also not be held so tightly that the fish's internal organs are damaged. Fish should never be held with dry hands. Overall, capture by net is probably the best method for removing fish from holding bins (in terms of comfort for both the operator and the fish). Fish should never be lifted or pulled by their tail or head as the stretching can inflame the spinal intercalary discs and cause dislocation of the vertebrae. Handling by the operculum can disrupt blood flow to the gills, or cause gill capillaries to haemorrhage, all of which seriously affect respiration.

Small fish are physically relatively easy to handle but can be particularly delicate. Larger fish can be more difficult to handle but can be held by supporting their caudal area (tail end of the body) and the ventral opercula regions (beneath the pectoral fins). Fish viscera are not well supported by mesenteries and muscle, and prolonged out-of-water handling increases the risk of internal injuries. For this reason, all fish should be kept in water in their natural horizontal position wherever, and for as long as, possible. Muscle strain on personnel should also be considered if very large (heavier than 10 kg) fish are being carried.

Carp sacks or weighing bags that can be kept wet are good methods of manipulating fish when they have to be removed from the water for weighing and so on.

When electric fishing and multiple fish are incapacitated, it is often tempting to hold fish in the net while repeatedly netting additional fish. This practice will markedly increase the adverse effects of the electric field on the fish and is a practice that should be discouraged.

Table 13.1 outlines measures that can be taken to minimize stress in fish.

13.2 Stress

Even minimal handling can result in an acute stress response in fish (elevated plasma cortisol levels etc.), which can take days or weeks from which to recover (Pickering *et al.* 1982). These stress responses can lead to reduced feeding (Pickering *et al.* 1982), and, if they reach high enough levels (e.g. due to descaling through rough handling or netting), it disrupts the fish's physiology and can lead to reduced disease resistance (Gadomski *et al.* 1994) and even reduced gamete viability. In extreme circumstances, it can also lead to the death of the fish.

Electric fishing has been widely reported as causing acute stress in fish (Mesa & Schreck 1989, Snyder 2003, Beaumont *et al.* 2000). Beaumont *et al.* (2000)

Table 13.1 Measures that can be taken to reduce stress during holding, handling and transportation of fish.

Problem	Suggested action	Comments
Duration of the stress response is usually proportional to duration of exposure.	Shorten the duration of stress.	Some effects may result in long recovery times.
Stress-induced mortality increases with water temperature.	Work at lower temperatures (e.g. use ice to cool water). Be aware of thermal stress caused by too low a temperature.	Not always practical under field conditions.
Stressors may be additive or synergistic.	Prevent simultaneous stress.	Possibly allow time between processes.
Abrasion between fish causes damage.	Reduce numbers handled per batch.	May conflict with time pressures.
Stress increases O_2 consumption, and ammonia and CO_2 output.	Use mild anaesthesia or sedation.	Note than anaesthetics themselves can act as stressors.

Adapted from Pickering (1993) and Ross and Ross (1999).

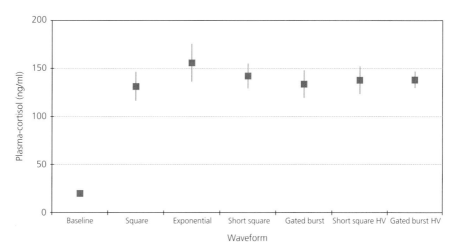

Figure 13.1 Mean (+/− 95%CL) blood plasma cortisol levels in rainbow trout pre and post shocking with a variety of pulsed direct current (pDC) waveforms (Beaumont *et al.* 2000).

found that blood plasma cortisol levels in rainbow trout were considerably elevated above baseline levels after electric fishing. Mean levels of electric fished samples were seven times those of the baseline samples. All individual waveforms caused significant increases over baseline levels, but no significant differences were found between the different waveforms evaluated (Figure 13.1). These stress responses from electric fishing are likely to be additive with subsequent exposure, may lead to behavioural changes in the fish (Mesa & Schreck 1989) and can affect the immune response system of the fish (VanderKooi *et al.* 2001). However, different species of fish may exhibit different responses (Davis & Parker 1986). Stress caused by handling can also influence survival, and Barrett and Grossman (1988) found handling stress was a greater determinant of survival than electric fishing in mottled sculpins.

13.3 Anaesthesia

Trying to control an active fish whilst handling it is likely to cause strong effects on both the fish's physiology and its subsequent behaviour (Tytler & Hawkins 1981). When fish are removed from water, individually or in groups, physiological stress is compounded by the risk of serious abrasion, mechanical shock and physical damage from operators constraining the fish. From a practical fisheries point of view, active fish also make measuring and weighing (especially on a pan balance) a difficult, frustrating and time-consuming process. By sedating the fish before handling, operators can reduce the adverse physiological reactions to handling stress, reduce the physical damage (especially to the scales and skin) that can be caused by the fish struggling (Ross & Ross 1999) and make processing faster and calmer for the fish (and personnel). However, anaesthetising fish is not without its own problems. Some anaesthetising chemicals can produce adverse effects (i.e. the fish react negatively to being put in the anaesthetic solution) (Readman *et al.* 2013), and anaesthetics procedures themselves can induce side effects that may not be considered desirable (e.g. reduced sperm mobility) (Allison 1961, quoted in Ross & Ross, 1999)).

The licences that many countries or organisations require to allow research on fish (e.g. the UK ASPA licence) specify the procedures that are allowed to be performed on the fish; this includes the use of anaesthetic agents. If a scientific procedure is considered to cause pain, suffering or distress, then anaesthesia should be used, and the administration of the anaesthesia itself may be a regulated procedure. It is against the spirit of the regulation and against principles of good fish welfare to withhold anaesthesia where it would normally be given just to avoid regulation.

An example of a scientific procedure that would require ASPA licensing under UK legislation is fin clipping of live fish for genetic analysis (with or without anaesthesia). Removal of a fin cannot be considered to be identification of an individual fish as all it does it split the population into marked (no fin) and unmarked (fin). It isn't possible, if further sampling is undertaken, to work out if it is the same individual fish that has been caught. Such fin removal for genetic analysis cannot therefore be considered as marking – it is a procedure to genetically test the animals caught to see how related they are.

Fin clipping is considered to cross the threshold for having the potential to cause pain, and it has the potential for infection; therefore, the protocol (under the ASPA licence) would need to include the use of anaesthesia.

Another example is gathering weight and length data under anaesthetic, unless it can be clearly shown that the weighing and measuring are always done in the particular group or population with the frequency planned for genuine husbandry reasons. Surveying a wild population is unlikely to come under the heading of husbandry.

Similarly, scale removal is performed frequently to evaluate year class. To be considered husbandry, the results should determine an action to take as part of a predefined management plan.

There are different levels of anaesthesia, and for most fishery applications full anaesthesia of the fish is not required. The level of anaesthesia does not, however, alter the decision regarding regulation (e.g. under ASPA). The lesser effect of sedation or tranquillisation (Stage I, Plane 1 anaesthesia (i.e. Stage I.1) in Table 13.2) can be induced either chemically or physically (by temperature reduction) to reduce physical activity whilst, for example, transporting fish over long distances. This will also reduce oxygen consumption (Taylor & Solomon 1979, Solomon & Hawkins 1981) and cause a reduction in the excretion of ammonia and carbon dioxide (Ferreira *et al.* 1984). Deeper sedation (Stage I.2) can enable easier handling of the fish whilst measuring, weighing and removing scales, thus preventing undue injury (McFarland 1959). More invasive procedures, such as peritoneal tag implantation, will require full surgical levels of anaesthesia (Stage II.2 or III in Table 13.2). Killing fish (euthanasia) can also be accomplished by very heavy concentrations of anaesthetic (Stage IV), although cerebral maceration is still recommended to confirm death.

Table 13.2 Classification of the behavioural changes that occur in fish during anaesthesia.

Level of anaesthesia			Behavioural responses
Stage	Plane	Description	
0		Normal (no anaesthesia)	Normal response to stimuli.
I	1	Light sedation	Response to external stimuli but activity reduced. Voluntary movement still possible.
	2	Deep sedation	No reaction to all but major stimuli. Some analgesia (pain suppression).
II	1	Light anaesthesia	Partial loss of equilibrium. Good analgesia
	2	Deep anaesthesia	Total loss of equilibrium (balance). Good analgesia.
III		Surgical anaesthesia	Total loss of reaction to even massive stimuli. Opercular rate very slow.
IV		Medullar (brain) collapse	Ventilation ceases, followed by cardiac arrest. Eventual death.

After McFarland (1959), Tytler and Hawkins (1981) and Ross and Ross (1999).

Several countries (including the United Kingdom, United States and Canada) only allow certain chemical compounds to be routinely used for anaesthetising fish (and other animals). Researchers can often, however, use other compounds under the terms of their regulatory license (e.g. ASPA). Lengthy withdrawal periods are often specified in the regulations before the fish can enter the human food chain; this includes not allowing release into the wild if the fish may be caught and eaten before the end of the withdrawal period. In the case of MS222, although the chemical is excreted and levels in the fish tissue decline to zero in 24 hours (Ross & Ross 1999), the US Food and Drug Administration (FDA) specify a 21-day withdrawal period.

Ross and Ross (1999) consider in depth the range of chemicals and techniques that can be used to anaesthetise and sedate aquatic animals, and the following will give only a brief overview of the principal chemical methods that can be used. Electroanaesthesia is covered in Chapter 12.

Anaesthetic compounds in general use in Europe, the United States and Canada are inhalation anaesthetics (i.e. the fish is placed in a solution of the drug, which it then absorbs through its gills during breathing). When the fish reaches the required level of sedation or anaesthesia, it is removed from the solution, processed and then placed in clean water to purge the compound from its system and thence recover. Every anaesthetic should be tested for the species being used and in the local water conditions. Some anaesthetics are better than others for relaxing and totally immobilising the fish (important if surgery or blood collection is taking place), and a chemical appropriate for the task should be used. The following are some of the substances commonly used for fish studies.

MS222 (ethyl-m-aminobenzoate, also known as tricaine methane sulfonate) is likely the most common anaesthetic agent used for fish and is the only anaesthetic drug approved by the US FDA, the UK Home Office and several other countries for use on fish destined for human consumption. It is registered for veterinary use with fish in the United States, Canada and most of Europe. The powder dissolves readily in water to provide a stock solution that can be further diluted for use. It should be kept in a dark container as it degrades in sunlight. It is also an acid and so should be buffered (using sodium bicarbonate) to reduce adverse effects of the acidic solution on the fish tissues (particularly if being used in low-pH waters). At the concentrations used for sedating fish, it is not toxic to humans but it still has a withdrawal period of 21 days.

Benzocaine (ethyl-4-aminobenzoate) is similar to MS222 in its effect upon fish but is only weakly soluble in water. For this reason, it is necessary to dissolve the stock solution in either acetone or alcohol prior to adding to water. A different formulation of benzocaine is benzocaine hydrochloride. It is soluble in water but very acidic in solution (unlike benzocaine which is neutral). Effective concentrations and lethal concentrations are similar to those found for MS222.

2-phenoxyethanol (2-PE) is a liquid that is easily dissolved in water. It is mildly bactericidal and antifungal which is an additional benefit if it is used for surgical procedures (tag implantation etc.). It has a wide safety margin, with effective concentrations for surgical anaesthesia being about 0.5 ml.l^{-1}. Lower concentrations (0.2 ml.l^{-1}) can be used to hold the fish in a low-sedation state for reasonable periods (5–10 min), making batch processing of fish simpler. It is rapidly excreted from fish, having a biological half-life of just 30 min. It is inexpensive compared with MS222, but in other respects it is similar to both MS222 and benzocaine.

Clove oil is increasingly being used in the United States as a fish anaesthetic. It has the principal advantage of having no known harmful effect on humans, and it is often used as a topically administered local anaesthetic for tooth pain. A derivative of this compound (Aqui-S) has been developed. This is a derivative of clove oil formulated specifically for use as a fish anaesthetic. It is reported to work well with little induction stress, albeit with a slow induction time (Ross & Ross 1999). In the United States, this compound is likely to be classified as 'safe' (i.e. for food use) by the US FDA and therefore will require no withdrawal period. The situation in the United Kingdom is as yet not certain as the compound has not yet received clearance for use by the European Medicine Evaluation Agency.

Finally, a brief mention will be made regarding using CO_2 to anaesthetise fish. Although there is some concern regarding the analgesia (pain suppression) that can be obtained with this chemical, it has been fairly widely used simply because of the ease in administering (injecting gaseous CO_2 into the water) and the possible lack of withdrawal needed. It is also possible to produce an anaesthetic solution by simply adding sodium bicarbonate or Alka Seltzer GOLD tablets (ones without aspirin) to the water. A dose rate of two to three tablets in 10–20 L of water has been found to be effective for juvenile steelhead and Chinook salmon. Dissolving CO_2 in the water will acidify it, and so suitable buffering of the water should take place. There is some debate in the United States as to whether this substance is outside the requirement for licensing by the FDA for use on food fish. Care is needed if working in a confined area, due to the toxic effect of the gas on humans.

It must be noted that, with the possible exception of CO_2, all the chemicals discussed here will also have Health & Safety implications for the users. For example, chronic exposure to 2-PE can impair fertility, and recent reports indicate that some retinal dysfunction can also occur after prolonged low-level exposure. Safety hazard data sheets of all chemicals used must be issued to users and the usual precautions taken, such as using eye protection and rubber gloves.

13.4 Fish density in holding bins

One of the most immediate effects of stress is the effect upon the respiratory system. Transported fish have been recorded as having an increased oxygen consumption of three times normal levels (Fröese 1985, 1986). In electric fishing, this stress is exacerbated by the fact that electrically narcotised fish will also have their breathing reduced. This may result in increased oxygen demand by the fish on recovery and also increase the concentrations of toxic respiratory end products that need to be excreted (Pickering 1993).

Some studies have been carried out regarding the transport of fish and their water and gas requirements, but few have examined short-term containment of fish. Conditions suitable for transport, however, can be considered as an optimum for the shorter term holding of fish.

Displacement of water can be used to calculate density of fish in a bin, and the specific gravity of fish (1.0) is roughly the same as for water. Therefore, a displacement of 1 litres of water equals a weight of ~1 kg fish. Under carefully controlled conditions, densities of up to 1 kg of fish per litre of water may be transported; however, densities of 50–350 g are more usual (Taylor & Solomon 1979). Some studies have been carried out based on the aquarium trade, detailing changes in water chemistry during plastic bag transport. Froese (1986), for example, gives a formula for the amount of fish that can be transported in oxygen (O_2)-saturated water.

To determine the oxygen loss from an unaerated bin, Beaumont *et al.* (2002) measured oxygen saturation in a container holding rainbow trout (being held prior to killing) on a local fish farm. The bin (a commercially available plastic dustbin) was half-filled with ~50 l of water, and the oxygen saturation measured. Over a 2 min period, approximately 20 kg of fish (56 fish, average weight 35 g) were added to the bin. This equates to about half the volume of water of fish. Oxygen saturation readings were taken at 2 min intervals until 5 min after adding fish; thereafter, readings were taken at 1 min intervals. Water temperature was 14° C. Results are shown in Figure 13.2 and show the speed with which O_2 levels fall before the rate of depletion slows, as the fish become more torpid due to lack of oxygen. The rate of depletion is temperature dependent and would decline faster at higher temperatures.

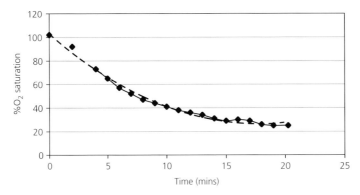

Figure 13.2 Depletion of oxygen in a bin after adding fish.

Keep-nets of some description provide a far better environment for holding fish. Those designed so that they resemble floating net cages are particularly good, provided there is enough water depth and the current is not too fast. Rigid boxes with mesh sides can also be placed in shallow areas of river for holding fish.

13.5 Oxygen and carbon dioxide

In order to mitigate against the detrimental effects, it is common practice to either aerate (add air) or oxygenate (add pure oxygen) the water in holding tanks. The purpose of this is to not only increase the dissolved oxygen (DO) concentration of the water, but also enable CO_2 removal (Taylor & Solomon 1979). For CO_2 to be removed from solution in the water, vigorous surface agitation is required. This aspect of CO_2 removal is easily overlooked. Winstone and Solomon (1976) described a situation where water saturated with O_2 built up toxic levels of CO_2 due to the low bubbling rate of the oxygen (and thus poor surface agitation).

The difference between DO concentrations low enough to cause mortality and those high enough for survival is small. Fish that appear healthy one minute may succumb very quickly if the DO concentration in their container falls. This may be due to water temperature increase or depletion due to respiration (Seager *et al*. 2000). Short-term exposure to sublethal low DO concentrations, however, is likely to result in minimal post-exposure mortality. Higher temperatures will require higher concentrations of oxygen due to its lower solubility at high temperatures (Table 13.3).

Aeration is a cheaper option than oxygenation but has the drawback of only adding 20% O_2 to the water. Aeration systems suitable for use in the field can range from simple zinc alkaline battery-operated units to powerful compressor systems powered from a vehicle electrical system. Cylinders of

Table 13.3 Temperature guidelines and suggested oxygen minimum limits – based on O_2 solubility data.

Dissolved O_2 saturation at different temperatures	
Temperature (°C)	Oxygen (mg.l^{-1})
5	12.8
10	11.3
15	10.2
20	9.2
25	8.2
30	7.5

compressed air can also be used, but apart from the cheaper cost of air compared with oxygen these negate many of the advantages that aeration has over the use of oxygen.

Oxygen is usually supplied from high-pressure (200 bar) cylinders but can be produced by chemical reaction (e.g. hydrogen peroxide breakdown; Taylor & Ross 1988) or by using liquid oxygen (Johnson 1979). Care needs to be taken regarding Health & Safety issues associated with using pure oxygen, and Health & Safety literature is available from gas suppliers. Carmichael *et al.* (1992) tabulated the efficiency of a range of oxygen diffuser systems (leaky pipe/micro-diffusers etc.) and concluded that all had efficiencies below 15%. They recommend that a combination of oxygen diffusion and surface aerators (to reduce CO_2 levels) be used.

Few studies have been carried out describing the specific changes in water quality (DO, CO_2 etc.) under real conditions. Fries *et al.* (1993) showed changes that occurred in a transport tank loaded with catfish, and Smith (1978) noted high O_2 consumption in the first hour for fish transport.

13.6 Ammonia

When holding fish in closed bins for any period of time, ammonia levels will rise. Ammonia is a waste product of fish metabolism and is excreted mostly via the gills. Ammonia dissolved in water changes into ammonium ions and hydroxyl ions. With increasing temperature and/or water acidity (pH), a proportion of the ammonia will dissociate into 'free ammonia' and water (Table 13.4). Free ammonia is considerably more toxic to fish than ammonium ions; salmonids, for example, show chronic reaction to concentrations as low as 0.006 mg.l⁻¹ of free ammonia. Decreases in the concentration of dissolved oxygen also increase the toxicity of the free ammonia (Merkens & Downing 1957, quoted in Solomon & Hawkins 1981).

Table 13.4 The maximum recommended level (mg.l⁻¹) of total ammonia (i.e. free ammonia **plus** ammonium ions for fish).

pH	Temperature		
	5 °C	15 °C	25 °C
6.5	50.0	22.2	11.1
7.0	16.7	7.4	3.6
7.5	5.1	2.3	1.2
8.0	1.6	0.7	0.4
8.5	0.5	0.3	0.1

13.7 **Temperature**

Temperatures that are lethal to fish will be species specific and will vary with acclimation (Taylor & Solomon 1979). Generally speaking, salmonids have a lower thermal tolerance than coarse fish, but little work has been carried out on determining individual species limits.

Scruton and Gibson (1995) working in cool-water Newfoundland streams considered that fishing for salmonids should not be carried out in water temperatures greater than 18 °C as mortalities were likely to occur.

Lowering ambient temperature can be used as a stress reduction method, and work has been carried out showing that slow cooling of fish can reduce the stress effects associated with transportation (Nakamura 1992).

13.8 **Osmotic balance**

Fish that have been damaged (e.g. mucous or scale removal) will have their osmoregulatory system disrupted. Depending upon whether the ionic concentration of the fish is higher or lower than its surrounding water, this will lead to either fluid loss or fluid gain. In most freshwaters the fish are likely to gain fluid, and in saltwater to lose water. In the case of the former, adding salt to the holding water, thus increasing the ionic concentration, can reduce fluid gain. Salt can also act as a fungal and disease inhibitor.

13.9 **Sensitive or robust fish**

Certain fish are renowned to be either very sensitive or very robust regarding the effect that electric fishing has on them. However, there is a wide range of differing perceptions about the category within which the same species falls. It is generally accepted that eels are probably the most robust and the salmonids (especially the adults) the most delicate. The gravidity status of the fish will also affect not only its likelihood of damage, due to changes in bone composition as the fish mature, but also the likelihood of damage and infertility of the eggs within the fish. It has been shown that a high cortisol (a stress by-product) level within fish affects both the future fertility of eggs and the viability of sperm, and Cho *et al.* (2002) found egg mortality of >90% in one-third of their test Chinook salmon. Electric fishing has also been shown to temporally adversely affect the immune response system in fish, leading to increased disease susceptibility post fishing (VanderKooi *et al.* 2001).

13.10 Fish eggs

As well as affecting fish eggs within the parent (see Section 13.9), electric fishing can also adversely affect fish eggs once they have been spawned (Snyder 2003). Harmful gradients for fry have been found to be around the threshold for causing harm to adult fish. Muth and Ruppert (1997) subjected fish eggs to voltage gradients of 1.2, 5 and 10 $V.cm^{-1}$. They found that all the voltage gradients could harm developing embryos. Bohl *et al.* (2010) found, for a range of species, vulnerability to electric shock–induced injury increased with increasing embryo size. However, Sternin *et al.* (1976) considered that only voltage gradients above 4–6 $V.cm^{-1}$ were damaging, and Roach (1999) considered that only eggs in very close proximity to the anode would experience high levels of mortality.

However, because of the possible risk of damage, at spawning periods areas of vegetation or gravel (i.e. salmonid redds) and other known spawning areas should be avoided when electric fishing (Lamarque 1990).

13.11 Bio-security

The ongoing spread of non-native aquatic pathogens and both aquatic and riparian non-native plants and animals has highlighted and increased the need for good bio-security measures and effective disinfecting of any equipment used in any water sampling. This is particularly pertinent for organisations or groups that may sample many different catchments or cover a wide geographic area, and many organisations are taking this issue very seriously (Figure 13.3).

Every vehicle should carry a disinfection box that should contain sufficient disinfection reagent, a spray system (small garden sprayers are ideal), a scrubbing brush and protective gloves. In the United Kingdom, Vircon Aquatic (DuPont) is commonly used as a disinfectant due to its ability to be safely disposed of near watercourses and its effectiveness against a wide range of pathogens.

After completion of fishing a site, all 'wet' gear should be cleaned and disinfected so that disease, parasites, aquatic organisms or plants are not transmitted between sites. Cleaning is not just a matter of rinsing muddy boots. ***All equipment used*** (boots, nets, bins, boats etc.) should be thoroughly cleaned of all mud and then sprayed or dipped in a suitable disinfectant solution. Boats should be thoroughly drained of water from the bilges and live wells and then flushed with a strong solution of disinfectant. Ideally, equipment should also be air-dried for 24 h after disinfecting. The reason for air-drying is that invertebrates have been found to be able to survive in the creases of some equipment for a period of time. If using stop nets that would be difficult to disinfect or air dry, the nets can be frozen for a few days in order to kill the organisms (3 days has been found to be sufficient for monofilament nets).

Figure 13.3 Inland Fisheries, Ireland, are taking the threat of invasive species very seriously.

When planning sampling within a catchment, sampling should proceed from the downstream site to the upstream site. This is to prevent non-native organisms or pathogens from being carried upstream when naturally they may not have been able to colonise upstream areas, or their spread upstream would have been very slow. This is particularly important if there are structures (e.g. dams, waterfalls or weirs) that would restrict the upstream movement of any natural carriers.

CHAPTER 14

Record keeping required

Comprehensive records should be kept of every electric fishing session. Although experimental research can provide guidelines, operators should carry out critical assessments of the effectiveness of their own settings (Reynolds & Holliman 2002). A field test kit for measuring voltage gradient (a penny probe plus oscilloscope/digital volt meter) should be available and staff trained in the equipment's use.

Records kept should include:

- Location, date, time of day, staff used and their role.
- The equipment used, the method (e.g. twin anode), the size of anode **and cathode** and the settings used when fishing. Record actual voltage gradients as shown by the field test kit.
- The environmental conditions such as weather, water conductivity and temperature.
- The reason for fishing should be recorded, including the target species and life stages. This is because differing techniques are likely to be used dependent upon the reason for the fish survey. If quick snap shot data are required (e.g. cyprinid species composition), then less efficient methodology will give a more cost-efficient result than the techniques that would be required for accurate numerical assessment of benthic species. When sampling different life stages or species, settings that are not suitable to other species or life stages may be more applicable (e.g. high frequencies for fry capture).
- Record the efficiency achieved.
- Record all occurrences of damage and/or death of fish. If possible, determine whether the reason for the damage and mortality is due to handling or electrical effect.

Electricity in Fish Research and Management: Theory and Practice, Second Edition. W.R.C. Beaumont.
© 2016 John Wiley & Sons, Ltd. Published 2016 by John Wiley & Sons, Ltd.

CHAPTER 15

Summary

The use of electricity, or electrical principles, has become a widespread and vital tool for managing and studying fish populations. By far the most common use of electricity is its use for capturing fish. This area is also where there is greatest likelihood of potential danger to staff and fish (due to wide variety of equipment and the close proximity of the operators to the current). A survey of electric fishing practices within two user groups revealed a great diversity of practice, a lack of consistency in approach to choice of equipment and settings, and varying levels of understanding of the basic principles of electric fishing (Lazauski & Malvestuto 1990, Beaumont *et al.* 2002). Many staff surveyed also reported occasional damage to fish by all methods.

The survey therefore demonstrated that there is a continuing need to develop and promote Best Practice Guidelines for electric fishing operations. Although many practitioners have good practical expertise in electric fishing, present levels of knowledge regarding theoretical aspects of electric fishing are very variable. All 'trained' personnel should have a good understanding of the basics of electric fishing, and one person in each team should have a comprehensive knowledge of the theory. This will allow intelligent adjustment of fishing parameters and a better understanding of factors influencing success or otherwise of different settings. In this way, settings can be used that will maximise fish welfare while still enabling efficient fishing operations to be carried out. Knowledge gained from the optimal settings found can be fed back into the system to improve overall understanding. With good knowledge of principles, standardisation of settings allowing better comparability of results is also achievable.

Once the basic theory and safe use of the equipment have been mastered, it is important to continue to learn from practice.

There has always been a concern for fish welfare while electric fishing, and this manual emphasises the concept of promoting fish welfare above fish capture. Information on basic electric circuit theory, choice of equipment, output characteristics and use should enable good fish capture efficiency with minimum incidence and severity of fish damage. Information and guidance that enable

Electricity in Fish Research and Management: Theory and Practice, Second Edition. W.R.C. Beaumont.
© 2016 John Wiley & Sons, Ltd. Published 2016 by John Wiley & Sons, Ltd.

users to have a good understanding of the factors that influence efficient equipment set-up and benign fish capture are fundamental to achieving these goals.

Recommending safe parameters to use when electric fishing in a wide range of conditions is difficult. Much of the literature on electric fishing, especially in respect of harmful effects on fish, is contradictory. However, notwithstanding the variable experiences of practitioners and inconsistencies in the published literature, it is possible to derive general principles for achieving optimum settings for electric fishing that minimise injury.

Whereas most research and guidance material seeks to achieve maximum fish capture, the guidance presented here suggests an alternative approach to electric fishing. This aims to use the most benign, rather than the most effective, electric fields to capture fish.

Temperature of water is the main criterion determining measures to maximise fish welfare, therefore electric

Finally, the advice of W.G. Hartley from around 1960 should be noted.

The best advice is to ensure that you know what you are talking about, check everything checkable on the electric side, and hope that as time goes on you will find fewer occasions on which you are surprised by the outcome of your actions. We all go on learning and only a fool will dare to dogmatize.

Glossary

General terms

Electric fishing The use of electric fields in water for the capture of fish, including the combined use of electric fields and mechanical methods.

Electric fishing apparatus The power supply, control gear, cables and electrodes used together for catching fish, or individual portions of a complete outfit.

Electric fish screen An electric field designed to repel or guide fish away from or around a barrier or area of danger.

Electroanaesthesia The use of electricity to induce short-term immobility in fish.

Electrocution of fish The killing of fish by means of electric current.

Electrofishing See *Electric fishing*.

Electro-immobilisation of fish The use of electricity for producing a temporary quiescent condition in fish.

Resistivity counter A fish counter based on the principle of detecting the change in electrical resistance of the water when a fish swims over an electrode array

Fishing methods

Backpack fishing Fishing using a self-contained electrical fishing machine carried on the back, while using the fishing electrodes. Power source can be a battery or small generator.

Boom-boat fishing Fishing from a boat fitted with fixed electrode arrays that are not manipulated by hand.

Electricity in Fish Research and Management: Theory and Practice, Second Edition. W.R.C. Beaumont.
© 2016 John Wiley & Sons, Ltd. Published 2016 by John Wiley & Sons, Ltd.

Classical electrical fishing	The use of a single or pair of hand-held electrodes connected to a generator. Operators wade in the water whilst operating the system.
Punt or boat fishing	A boat in which operators stand and from which fishing is carried out using hand-held electrodes.
Tote-boat fishing	A small boat used to carry equipment but not personnel while fishing.

Equipment

Anode	An electrode with a positive potential relative to earth.
Cathode	An electrode with a negative potential relative to earth.
Dead-man's switch	Switch on anode or in anode circuit (either hand-held or boom-mounted) that requires constant pressure for the electrodes to be energised.
Dropper or pendant	Rod or tube electrodes that hang from a supporting frame (e.g. fish screen electrodes or Wisconsin ring).
Electrode array	A pattern of electrodes arranged in a distinct conformation to produce a defined electric field in water.
Emergency off switch	Switch that cuts off electrical supply to electrodes and/ or pulse box when hit.
Generator	Machine designed to produce electricity.
Pulse box	Box containing circuitry required to modify generator output to that suitable for electric fishing.
Ring or torus electrode	An electrode consisting of a metal hoop (which may be other than circular in plan).
Rod or tube electrode	A cylindrical electrode with a length that is great with respect to its diameter.
Sphere electrode	A metal sphere used as an electrode.
Wisconsin ring electrode	A circular frame or other arrangement from which is hung a series of pendant or dropper electrodes so spaced as to have the electrical characteristics of an electrode of the diameter of the complete dropper ring.

Electrical terms

Alternating current (polyphase)	A series of equal AC waveforms of the same frequency but displaced in phase in a uniform sequential rhythm.

Alternating current	A current which varies sinusoidally between equal positive (single phase) and negative values at a uniform frequency.
Ambient conductivity	As for specific conductivity but not temperature corrected (i.e. the value at the ambient temperature).
Anode field	The space enclosing an anode system in which a potential gradient due to that anode system can be detected.
Cathode field	The space enclosing a cathode system in which a potential gradient due to that cathode system can be detected.
Conductivity	The ability of a material to conduct electric charge (the reciprocal of resistivity), measured in Siemens.
Current density	The local value of electric current carried by a unit area perpendicular to the current lines. Usually expressed as amperes per square centimetre.
Current lines	Lines perpendicular to the equipotential lines, which indicate the instantaneous direction in which the electric current flows.
Cycle	The complete sequence intervening between two successive corresponding points in a regularly recurring sequence of potential variations (i.e. a periodic voltage waveform).
Decay slope	The part of a single pulse included between its steady or peak value and its defined end.
Direct current	A current resulting from the discharge of a uniform potential through a circuit with constant properties.
Duty cycle	The percentage time within one cycle where current is flowing.
Effective zone	The area within which an electrode produces a compulsive effect on the fish encountering it.
Efficiency	The percentage of fish caught by the electrode system. It can be expressed in term of total population or single species.
Electrode field	The zone surrounding an electrode within which its potential gradient can be readily detected.
Equipotential lines	Lines joining points in an electric field which have at simultaneous instants potential values which are equal.
Exponential pulsed current	The current pattern resulting from the complete discharge of a capacitor through a conductor, repeated at equal intervals to produce a uniform series.
Field pattern	The distribution of potential and current in an electric conduction field.

Frequency	The number of complete oscillations executed by a periodic alternating voltage in unit time. The standard unit is the Hertz, representing one cycle per second.
Full-wave rectified DC	The unidirectional current derived from reversing the polarity of either the +ve or −ve component of an alternating current in such a way as to produce a series of adjacent symmetrical waves of the same polarity at a frequency double that of the original AC.
Half-wave rectified DC	The unidirectional current derived from suppressing either the +ve or −ve component of an alternating current, leaving a series of disconnected waves of the same polarity, with the same frequency as the original AC.
Heterogeneous field	An electrical field in which the current density and voltage gradient are not uniform in space and (or) time.
Homogeneous field	An electrical field in which the current density and voltage gradient are uniform.
Mean voltage	The arithmetic mean, or average, value measured over an integral number of complete cycles of a periodic voltage waveform. Similar to, but not the same as, root mean square (RMS) voltage.
On/off time ratio	The ratio of pulse duration to pause duration.
Pause duration	The period between the defined end of one pulse and the start of the next.
Peak-to-peak voltage	The magnitude of the minimum to maximum instantaneous voltage appearing between the electrodes.
Peak voltage	The magnitude of the zero to maximum instantaneous voltage appearing between the electrodes.
Period	The duration of a single cycle.
Potential gradient *or* **voltage gradient**	The difference of potential measured over a stated distance. Usually given in volts per centimetre. The normal metric for voltage field intensity.
Power transfer theory	Theory stating that fish reaction to an electric field is related to the product of the voltage gradient plus the current density, and varies depending on the ratio of the fish and water conductivity.
Pulsed current	A current consisting of uniform discrete discharges, in a regular sequence.
Pulse duration	This depends on the shape of the pulse. A. Square wave: the duration of current flow. B. Exponential pulse: the period between Vmax and V/e, where e is the base of natural logarithm. C. With sinusoidal pulses, the duration has been taken as the period during which the potential exceeds 10% of the peak value.

Quarter sine wave	The unidirectional current pattern obtained when a rectified DC is switched so that it has a vertical rise between zero and Vpeak.
Resistivity	A measure of the ability of a substance to oppose the flow of electrical charge, measured in Ohms.
Ripple	The residual cyclic variation in voltage (peak-to-peak voltage) in a smoothed rectified DC.
Rise slope	The part of a single pulse included between its commencement and its steady or peak value.
RMS voltage	The root mean square value of a periodic voltage waveform. This is equivalent to the value of a steady DC voltage that would dissipate the same mean power in the same resistive load.
Sawtoothed current	Pulses of current produced by a potential where the rise slope is linear and the fall slope is vertical, or vice versa.
Smoothed rectified DC	A rectified AC current in which the cyclic variation in peak-to-peak voltage is reduced.
Specific conductivity	The conductivity of a material at a standard temperature (commonly 25°C). The value is commonly expressed in Siemens per centimetre.
Square wave pulsed current	The current pattern resulting from a uniform series of a constant DC voltage.
Unidirectional current	A current produced by an electric potential that may vary or be interrupted, but is never reversed.

References

Alabaster, J.S. & Hartley, W.G. (1962). The efficiency of a direct current electric fishing method in trout streams. *Journal of Animal Ecology* **31**, 385–388.

Allard, L., Grenouillet, G., Khazraie, K., Tudesque, L., Vigouroux, R. & Brosse, S. (2014). Electrofishing efficiency in low conductivity neotropical streams: towards a non-destructive fish sampling method. *Fisheries Management and Ecology* **21**, 234–243.

Allen-Gil, S.M. (Ed.) (2000). New perspectives in electrofishing. Report to US Environmental Protection Agency EPA/600/R-99/108.

Andrus, C. (2000). Experiences from the field: alcove sampling on the Willamette River, OR. In: *New Perspectives in Electrofishing* (Ed. S. Allen-Gil.) Report to US Environmental Protection Agency EPA/600/R-99/108.

Anonymous (1918). *The Encyclopedia Americana: A Library of Universal Knowledge.* Encyclopedia Americana, New York.

Anonymous (2003a). Water quality – Sampling of fish with electricity. *CEN/CENELECT* **14011**, 2003.

Anonymous (2003b). Household and similar electrical appliances – Safety – Part 2-86: Particular requirements for electric fishing machines. IEC 60335-2-86 Ed.2.0.

Anonymous (2009). *Electrical Safety: Safety and Health for Electrical Trades.* DHHS (NIOSH) Publication Number 2009–113 April 2009. Department of Health and Human Services, Centers for Disease Control and Prevention, National Institute for Occupational Safety & Health, Washington, DC.

Baggs, I. (1863). Paralysing Fish Birds etc. Patent No 2644. H.M. Stationary Office, London.

Bain, M.B., Finn, J.T. & Booke, H.E. (1985). A quantitative method for sampling riverine microhabitats by electrofishing. *North American Journal of Fisheries Management* **5**, 489–493.

Balachandran, A., Krishnan, B. & John, L. (2013). Accidental deaths due to electrocution during electro-fishing. *Journal of Evolution of Medical and Dental Sciences* **2**, 9376–9379.

Baldwin, L. & Aprahamian, M. (2012). An evaluation of electric fishing for assessment of resident eel in rivers. *Fisheries Research* **123–124**, 4–8. doi:10.1016/j.fishres.2011.11.011

Baras, E. (1995). An improved electrofishing methodology for the assessment of habitat use by young-of-the-year fishes. *Archiv für Hydrobiologie* **134**, 403–415.

Barrett, J.C. & Grossman, G.D. (1988). Effects of direct current electrofishing on the mottled sculpin. *North American Journal of Fisheries Management* **8**, 112–116.

Bary, B.M. (1956). The effect of electric fields on marine fishes. *Marine Research Series* **1**, 36.

Bateman, D.S., Gresswell, R.E. & Torgersen, C.E. (2005). Evaluating single-pass catch as a tool for identifying spatial pattern in fish distribution. *Journal of Freshwater Ecology* **20**, 2, 335–345.

Electricity in Fish Research and Management: Theory and Practice, Second Edition. W.R.C. Beaumont.
© 2016 John Wiley & Sons, Ltd. Published 2016 by John Wiley & Sons, Ltd.

Bayley, P.B., Larimore, R.W. & Dowling, D.C (1989). Electric seine as a fish-sampling gear in streams. *Transactions of the American Fisheries Society* **118**, 447–453.

Beaumont, W.R.C., Mills, C.A. & Williams, G. (1986). The use of a microcomputer as an aid to identifying objects passing through a resistivity fish counter. *Aquaculture and Fisheries Management* **17**, 213–226.

Beaumont, W.R.C., Lee, M.J. & Rouen, M. (1997). *Development of Lightweight Backpack Electric Fishing Gear*. Report to the Environment Agency. Environment Agency, Swindon.

Beaumont, W.R.C., Lee, M. J. & Rouen, M. A. (1999). *Development of lightweight Backpack Electric Fishing Gear – Phase II*. Final Report to Environment Agency (National Coarse Fish Centre). Environment Agency, Swindon.

Beaumont, W.R.C., Lee, M. & Rouen, M.A. (2000). An evaluation of some electrical waveforms and voltages used for electric fishing; with special reference to their use in backpack electric fishing gear. *Journal of Fish Biology* **57**, 433–445.

Beaumont, W.R.C., Taylor, A.A.L., Lee, M.J. & Welton, J.S. (2002). *Guidelines for Electric Fishing Best Practice*. R&D Technical Report W2-054/TR. Environment Agency, Swindon.

Beaumont, W.R.C., Lee, M.J. & Peirson, G (2003). *An Investigation of the Equivalent Resistance, Power Requirements and Field Characteristics of Electric Fishing Electrodes*. Report to Environment Agency, EA Technical Report W2-076. Environment Agency, Swindon.

Beaumont, W.R.C., Lee, M.J., & Peirson, G. (2005). The Equivalent Resistance and Power Requirements of Electric Fishing Electrodes. *Fisheries Management & Ecology* **12**, 37–44.

Beaumont, W.R.C. (2005). Factors Affecting Electric Fishing Best Practice. In: *Proceedings of the Institute of Fisheries Management 34th Annual Study Course*, 49–65, Nottingham.

Beaumont, W.R.C., Pinder, A.C., Scott L. & Ibbotson A.T. (2007). A history of the river Frome salmon monitoring facility and new insights into the ecology of Atlantic salmon. In: *Proceedings of the Institute of Fisheries Management 37th Annual Study Course*, Minehead, 179–194.

Bird, D.J. & Cowx, I.G. (1993). The selection of suitable pulsed electric currents for electric fishing. *Fisheries Research* **18**, 363–376.

Blasius & Schweitzer, F (1893). Beschreibung u. Benennung der "Galvano-Narcose": Beobachtung der "Galvano-Hypnose". *Pflüger's Archive* **53**, 527.

Boccardy J.A. & Cooper E.L. (1963). The use of rotenone and electrofishing in surveying small streams. *Transactions of the American Fisheries Society* **92**, 3, 307–310.

Bodrova, N.V. & Krayukhin, B.V. (1959). K voprosu o "vidovoi" chuvstvitel 'nosti ryb k elektricheskomu toku (The problem of the of the "specific" sensitivity of fish to electric current) *Byullten Instituta Biologii Vodokhranilishch* No. 5.

Bohl, R.J., Henry, T.B. & Strange R.J. (2010). Electroshock-induced mortality in freshwater fish embryos increases with embryo diameter: a model based on results from 10 species. *Journal of Fish Biology* **76**, 975–986.

Bohlin, T. (1982). The validity of the removal method for small populations – consequences for electric fishing practice. *Report of Institute of Freshwater Research*, Drottningholm **60**, 15–18.

Bohlin, T. (1989). Electrofishing – theory and practice with special emphasis on salmonids. *Hydrobiologia* **173**, 9–43.

Borgstroem, R. & Skaala, O. (1993). Size-dependent catchability of brown trout and Atlantic salmon parr by electrofishing in a low conductivity stream. *Nordic Journal of Freshwater Research*. Drottningholm **68**, 14–20.

Borkholder, B.D & Parsons, B.G. (2001). Relationship between electrofishing catch rates of age-0 walleyes and water temperature in Minnesota lakes. *North American Journal of Fisheries Management* **21**, 318–325.

Bouck, G. & Ball, R. (1966). Influence of capture methods on blood characteristics and mortality in rainbow trout. *Salmo gairdneri. Transactions of the American Fisheries Society* **95**, 170–176.

Bourgeois C E (1995). Electrofishing techniques employed by the Enhancement and Aquaculture section in determining the effectiveness of fry stocking. In: Scruton, D.A. & Gibson, R.J. (Ed.), *Canadian Manuscript Report of Fisheries & Aquatic Science*. 83–96.

Bowles, F.J., Frake, A.A. & Mann, R.H.K. (1990). A comparison of efficiency between two electric fishing techniques on a section of the River Avon, Hampshire. In: I.G. Cowx (ed.) *Developments in Electric Fishing*, 229–235. Oxford: Fishing News Books, Blackwell Scientific Publications.

Bozek, M.A. & Rahel, F.J. (1991). Comparison of streamside visual counts to electrofishing estimates of Colorado River cutthroat trout fry and adults. *North American Journal of Fisheries Management* **11**, 38–42.

Brobbel, M.A., Wilkie, M.P., Davidson, K., Kieffer, J.D., Bielak, A.T. & Tufts, B.L. (1996). Physiological effects of catch and release angling in Atlantic salmon (*Salmo salar*) at different stages of freshwater migration. *Canadian Journal of Fisheries and Aquatic Sciences* **53**, 2036–2043.

Brøther, D. (1954). Electric fishing. *Teknisk Ukeblad* **101**, 369–376.

Brun, M., Abraham, Jarry, M. Dumas, J., Lange, F. & Prévost, E. (2010). Estimating an homogeneous series of a population abundance indicator despite changes in data collection procedure: a hierarchical Bayesian modelling approach. *Ecological Modelling* **222** (2011), 1069–1079.

Burkhardt, R.W. & Gutreuter, S. (1995). Improving electrofishing catch consistency by standardizing power. *North American Journal of Fisheries Management* **15**, 375–381.

Burrows, R.E. (1957). Diversion of adult salmon by an electrical field. *US Fish and Wildlife Service Special Scientific Report on Fisheries*. No. 246.

Bussell, R.B. (1978). Fish Counting Stations: Notes for guidance in their design and use. *Department of the Environment Report*, London. November 1978: 68pp + appx.

Carle, F.L. & Strub, M.R. (1978). A new method for estimating population size from removal data. *Biometrics* **34**, 621–630.

Carmichael, G.J., Jones, R.M. & Morrow, J.C. (1992). Comparative efficacy of oxygen diffusers in a fish-hauling tank. *The Progressive Fish-Culturist* **54**, 35–40.

Carrier, P., Rosenfeld, J. & Johnson, R.M. (2009). Dual-gear approach for calibrating electric fishing capture efficiency and abundance estimates. *Fisheries Management and Ecology* **16**(2), 139–146.

Cave, J. (1990). Trapping salmon with the elecronet. In I. Cowx (ed.), Developments in electric fishing. *Proceedings of an International Symposium on Fishing with Electricity*, Hull, UK, 1988. Oxford: Fishing News Books, Blackwell Scientific Publications. Pages 65–70.

Chaput, G., Moore, D. & Peterson, D. (2005). Predicting Atlantic salmon (*Salmo salar*) juvenile densities using catch per unit effort open site electrofishing. *Canadian Technical Report of Fisheries and Aquatic Science* No. 2600. 25 pp.

Chick, J.H., Coyne, S. & Trexler, J.C. (1999). Effectiveness of airboat electrofishing for sampling fishes in shallow, vegetated habitats. *North American Journal of Fisheries Management* **19**, 957–967.

Cho, G.K., Heath, J.W. & Heath, D.D. (2002). Electroshocking influences chinook salmon egg survival and juvenile physiology and immunology. *Transactions of the American Fisheries Society* **131**, 224–233.

Collins, G.B., Volz, C.D. & Trefethen, P.S. (1954). Mortality of salmon fingerlings exposed to pulsating direct current. *Fishery Bulletin of the Fish and Wildlife Service. U.S.*, 56, 61–81.

Cooke, S.J., Bunt, C.M. & McKinley, R.S. (1998). Injury and short term mortality of benthic stream fishes – a comparison of collections techniques. *Hydrobiologia* **379**, 207–211.

Copp, G.H. (1989). Electrofishing for fish larvae and 0+ juveniles: equipment modifications for increased efficiency with short fishes. *Aquaculture and Fisheries Management* **20**, 453–462.

Corcoran, M.F. (1979). Electrofishing for catfish: use of low-frequency pulsed direct current. *The Progressive Fish-Culturist* **41**, 200–201.

Cowx, I.G. (1983). Review of the methods for estimating fish population size from removal data. *Fisheries Management* **14**, 67–82.

Cowx, I.G. (ed.) (1990). *Developments in Electric Fishing*, 34–40. Oxford: Fishing News Books, Blackwell Scientific Publications.

Cowx, I.G., Wheatley, G.A., Hickley, P. & Starkie, A.S. (1990). Evaluation of electric fishing equipment for stock assessment in large rivers and canals in the United Kingdom. In: I.G. Cowx (ed.) *Developments in Electric Fishing*, 34–40. Oxford: Fishing News Books, Blackwell Scientific Publications.

Crisp, D.T. & Crisp, D.C. (2006). Problems with timed electric fishing assessment methods. *Fisheries management and Ecology: Management Note* **13**, 211–212.

Cross, D.G. & Stott, B. (1975). The effect of electric fishing on the subsequent capture of fish. *Journal of Fish Biology* **7**, 349–357.

Crozier, W.W. & Kennedy, G.J.A. (1994). Application of semi-quantitative electrofishing to juvenile salmonid stock surveys. *Journal of Fish Biology* **45**, 159–164.

Cuinat, R. (1967). Contribution to the study of physical parameters in electrical fishing in rivers with direct current. In: Vibert, R. (ed.), *Fishing with Electricity: Its Application to Biology and Management*: 131–171. Fishing News Books Limited, Farnham, Surrey.

Cunjak, R.A., Randall, R.G. & Chadwick, E.M.P. (1988). Snorkeling versus electrofishing: a comparison of census techniques in Atlantic salmon rivers. *Naturaliste Canadien* **115**, 89–93.

Cunningham, K.K. (1998). Influence of environmental variables on flathead catfish electrofishing catch. *Proceedings of the Annual Conference of the Southeastern Association of Fish and Wildlife Agencies* **52**, 125–135.

Dalbey, S.R., Mahon, T.E. & Fredenberg, W. (1996). Effect of electrofishing pulse shape and electrofishing-induced spinal injury on long-term growth and survival of rainbow trout. *North American Journal of Fisheries Management* **16**, 560–569.

Danyulite, G.P & Prits, K.I. (1965). The reaction of fish to pulsating electric current *VOP IKHTIOL* **5**, 338–346

Danyulite, G.O. & Malyukina, G.N. (1967). Issledovanie fiziologicheskogo mekhanizma deistviya polya postoyannogo elektricheskogo toka na ryb. *Povedenie I retseptsii ryb. Izdatel'stvo "Nauka"*

Dauphin, G., Prevost, E., Adams, C.E. & Boylan, P. (2009). A Bayesian approach to estimating Atlantic salmon fry densities using a rapid sampling technique. *Fisheries Management and Ecology* **16**, 399–408

Davidson, P.N. (1984). The Use of Electricity for Fish Capture: The Repsonse of Some Freshwater Fish to Frequencies of Pulsed Direct Current. Unpublished MSc. dissertation. University of Wales, Newport.

Davis, K.B. & Parker, N.C. (1986). Plasma corticosteroid stress response of fourteen species of warm-water fish to transportation. *Transactions of the American Fisheries Society* **115**, 499.

De Lury, D.B. (1947). On the estimation of biological populations. *Biometrics* **3**, 145–167.

Dewey, M.R. (1992). Effectiveness of a drop net, a pop net, and an electrofishing frame for collecting quantitative samples of juvenile fishes in vegetation. *North American Journal of Fisheries Management* **12**, 808–813.

Dolan, C.R. & Miranda, L.E. (2003). Immobilisation Thresholds and Electrofishing Relative to Fish Size. *Transactions of the American Fisheries Society* **132**, 969–976.

Dumont, S.C. & Dennis, J.A. (1997). Comparison of day and night electrofishing in Texas reservoirs. *North American Journal of Fisheries Management* **17**, 939–946.

Edwards, J.L. & Higgins, J.D. (1973). The effects of electric currents on fish. *Engineering Experimental Station, Georgia Institute of Technology Final Technical Report* B-397, B-400, E-200-301, 15pp.

Edwards, M.R., Combs, D.L., Cook, S.B. & Allen, M. (2003). Comparison of single-pass electro-fishing to depletion sampling for surveying fish assemblages in small warmwater streams. *Journal of Freshwater Ecology* **18**, 4, 625–634.

Ferguson, R.A. & Tufts, B.L. (1992). Physiological effects of brief air exposure in exhaustively exercised rainbow trout (*Oncorhynchus mykiss*): implications for "catch and release" fisheries. *Canadian Journal of Fisheries and Aquatic Sciences* **49**, 1157–1162.

Ferreira, J.T., Schoonbee, H.J. & Smit, G.L. (1984). The use of benzocaine-hydrochloride as an aid in the transport of fish. *Aquaculture* **42**, 169–174.

Fisher, W.L. & Brown, M.E. (1993). A prepositioned areal electrofishing apparatus for sampling stream habitats. *North American Journal of Fisheries Management* **13**, 807–816.

Flux, C. (1967). Factors affecting the response of trout to an electtric field in fresh and salt water. *Journal of the Fisheries Research Board of Canada* **24**, 191–199.

Fries, J.N., Berkhouse, C.S., Morrow, J.C. & Carmichael, G.J. (1993). Evaluation of an aeration system in a loaded fish-hauling tank. *The Progressive Fish-Culturist* **55**, 187–190.

Fröese, R. (1985). Improved fish transport in plastic bags. *ICLARM (International Center for Living Aquatic Resources Management) Newsletter* **8**, 8–9.

Fröese, R. (1986). How to transport live fish in plastic bags. *Infofish Marketing Digest* **86**, 35–36.

Funk, J.L. (1949). Wider application of the electrical method of collecting fish. *Transactions of the American Fisheries Society* **77**, 1, 49–60.

Gadomski, D.M., Mesa, M.G. & Olson, T.M. (1994). Vulnerability to predation and physiological stress responses of experimentally descaled juvenile Chinook salmon *Oncorhynchus tshawytscha*. *Environmental Biology of Fishes* **39**, 191–199.

Gargan, P.G., Stafford, T., Økland, F. & Thorstad E.B. (2015). Survival of wild Atlantic salmon (*Salmo salar*) after catch and release angling in three Irish rivers. *Fisheries Research* **161**, 252–260.

Garner, P. (1997). Sample sizes for length and density estimation of 0+ fish when using point sampling by electrofishing. *Journal of Fish Biology* **50**, 95–106.

Gerking, S.D. (ed.) (1978). *Ecology of Freshwater Fish Production*. Blackwell Scientific Productions.

Goodchild, G.A. (1990). Electric Fishing and Safety. In: I.G. Cowx and Lamarque (Ed.) *Fishing with Electricity. Application in Freshwater Fisheries Management*. Fishing News Books, Blackwell Scientific Publications. Pages 157–175.

Gosset, C. & Rives, J. (2005). Anesthesie et procedtes chirurgicales pour l'implantation de radio emetteurs dan la cavite de truites communes adults (Salmo trutta). *Bulletin Francaise de la Pesche et de la Pisciculture* **374**, 21–34.

Growns, I.O., Pollard, D.A. & Harris, J.H. (1996). A comparison of electric fishing and gillnetting to examine the effects of anthropogenic disturbance on riverine fish communities. *Fisheries Management and Ecology* **3**, 13–24.

Hadderingh, R.H. & Jansen, H. (1990). Electric fish screen experiments under laboratory and field conditions. I.G. Cowx (ed.), *Developments in Electric Fishing*. Oxford: Fishing News Books, Blackwell Scientific Publications. Pages 266–280.

Halsband, E. (1967). Basic principles of electric fishing. In: Vibert, R. (ed.) *Fishing with Electricity: Its Application to Biology and Management*. Fishing News Books Limited, Farnham, Surrey. Pages 57–64.

Hardin, S. & Connor, L.L. (1992). Variability of electrofishing crew efficiency, and sampling requirements for estimating reliable catch rates. *North American Journal of Fisheries Management* **12**, 612–617.

Hartley, W.G. (1975). Electrical fishing apparatus and its safety. *Fisheries Management* **6**, 73–77.

Hartley, W.G. (1980a). The use of electrical fishing for estimating stocks of freshwater fish. In: Backiel, T. & Welcomme, R.L. (eds) *Guidelines for Sampling Fish in Inland Waters*: 91–95. EIFAC Technical Paper No. 33 (European Inland Fisheries Advisory Commission).

Hartley, W.G. (1980b). The Use of Three-Phase Current for Electrical Fishing. *Fisheries Management* **11**, 2, 77–79.

Haskell, D.C., MacDougal, J. & Geduldig, D. (1954). Reactions and motion of fish in a direct current electric field. *New York Fish and Game Journal* **1**, 47–64.

Hayes, J.W. & Baird, D.B. (1994). Estimating relative abundance of juvenile brown trout in rivers by underwater census and electrofishing. *New Zealand Journal of Marine and Freshwater Research* **28**, 243–253.

Heggenes, J., Braband, A. & Saltveit, S.J. (1990). Comparisons of three methods for studies of stream habitat use by young brown trout and Atlantic salmon. *Transactions of the American Fisheries Society* **119**, 101–111.

Heidinger, R.C., Helms, D.R., Hiebert, T.I. & Howe, P.H. (1983). Oerational comparison of three electrofishing systems. *North American Journal of Fisheries Management* **3**, 254–257.

Hellawell, J.M. (1973). Automatic methods of monitoring salmon populations. *International Atlantic salmon symposium, St Andrews, Canada* **4**, 317–337.

Hellawell, J.M., Leatham, H. & Williams, G.I. (1974). The upstream migratory behaviour of salmonids in the river Frome, Dorset. *Journal of Fish Biology* **6**, 729–744.

Henry, T.B. & Grizzle, J.M. (2006). Electric-induced mortality of newly transformed juvenile fishes in waters of different conductivity. *Journal of Fish Biology* **68**, 747–758.

Herman, L. (1885). Eine Wirkung galvanische Strome auf Organismen. *Pflüger's Archiv. Für die ges. Physiol.* **37**, 457–460.

Herman, L. & Matthias, Fr. (1886). Der Galvanotropismus der larven von *Rana* temporaria und der Fische. *Pflüger's Archiv. Für die ges. Physiol.* **57**, 391–405.

Hickley, P. (1985). Electric Fishing. *Institute of Fisheries Management Advisory Leaflet* 21pp.

Hickley, P. (1990). Electric Fishing in Practice. In: Cowx, I.G. & Lamarque, P. (eds.) *Fishing with Electricity: Applications in Freshwater Fisheries Management* Fishing News Books, Blackwell Scientific Publications. Pages 176–185.

Hickley, P. & Millwood, B. (1990). the UK safety guidelines for electric fishing: its relevance and application. In: I.G. Cowx (ed.) *Developments in Electric Fishing*.34–40 Oxford: Fishing News Books, Blackwell Scientific Publications. Pages 311–323

Hill, T.D. & Willis, D.W. (1994). Influence of water conductivity on pulsed-AC and pulsed-DC electrofishing catch rates for largemouth bass. *North American Journal of Fisheries Management* **14**, 202–207.

Hollander, B.A. & Carline, R.F. (1994). Injury of wild brook trout by backpack electrofishing. *North American Journal of Fisheries Management* **14**, 643–649.

Holzer, W. (1932). Der elektrische Fischrechen. *Fischerie Zeitung* **35**, 218.

Hughes, R.M., Kaufmann, P.R., Herlihy, A.T., Intelmann, S.S., Corbett, S.C., Arbougast, M.C. & Hjort, R.C. (2002). Electrofishing distance needed to estimate fish species richness in raftable Oregon rivers. *North American Journal of Fisheries Management* **22**, 1229–1240.

James, P.W., Leon, S.C., Zale, A.V. & Maughan, O.E. (1987). Diver-operated electrofishing device. *North American Journal of Fisheries Management* **7**, 597–598.

Janac, M. & Jurajda, P. (2005). Inter-calibration of three electric fishing techniques to estimate 0+ juvenile fish densities on sandy river beaches. *Fisheries Management and Ecology* **12**, 3 161–167.

Jesien, R.V. & Horcutt, C.H. (1990). Method for evaluating fish response to electric fields. In: Cowx, I.G. (ed.) *Developments in Electric Fishing*: Fishing News Books, Blackwell Scientific Publications. Pages 10–18.

Johnson, I.K., Beaumont, W.R.C. & Welton, J.S. (1990). The use of electric fish screens in the Hampshire Test and Itchen, England. In: Cowx, I.G. (ed.) *Developments in Electric Fishing*: Fishing News Books, Blackwell Scientific Publications. Pages 256–265.

Johnson, S.K. (1979). Transport of live fish. *Aquaculture Magazine* **5**, 20–24.

Justus, B. (1994). Observations on electrofishing techniques for three catfish species in Mississippi. *Proceedings of the Annual Conference of the Southeastern Association of Fish and Wildlife Agencies* **48**, 524–532.

Kennedy, G.J.A. & Strange, C.D. (1981). Efficiency of electric fishing for salmonids in relation to river width. *Fisheries Management* **12**, 55–60.

Knights, B., Bark, A., Ball, M., Williams, F., Winter, E. & Dunn, S. (2001). Eel and elever stocks in England and Wales – Status and management options. *Environment Agency Technical Report* W248, Bristol

Kocovsky, P.M., Gowan, C., Fausch, K.D. & Riley, S.C. (1997). Spinal injury rates in three wild trout populations in Colorado afer eight years of backpack electrofishing. *North American Journal of Fisheries Management* **17**, 308–313.

Kolz, A.L. (1989). A power transfer theory for electrofishing. In: *Electrofishing, a Power Related Phenomenon*: 1–11. *Fish and Wildlife Technical Report 22. United States Department of the Interior*, Fish and Wildlife Service, Washington, DC.

Kolz, A.L. & Reynolds, J.B. (1989). Determination of power threshold response curves. In: *Electrofishing, a Power Related Phenomenon*: 15–24. Fish and Wildlife Technical Report 22. United States Department of the Interior, Fish and Wildlife Service, Washington, DC.

Kolz, A.L. & Reynolds, J.B. (1990). A power threshold method for the estimation of fish conductivity. In: I.G. Cowx (ed.) *Developments in Electric Fishing*. Oxford: Fishing News Books, Blackwell Scientific Publications. Pages 5–9.

Kolz, A.L. (1993). In-water Electrical Measurements for Evaluating Electrofishing. *US Fish and Wildlife service. Fish and Wildlife Technical Report* 22 pp.

Kolz, A.L., Reynolds, J., Temple, A., Boardman, J. & Lam, D. (1998). Principles and Techniques of Electrofishing (course manual). *Branch of Aquatic Resources Training, U.S. Fish and Wildlife Service, National Conservation Training Center*, Shepherdstown, West Virginia.

Kolz, A.L. (2006). Electrical conductivity as applied to electrofishing. *North American Journal of Fisheries Management* **135**, 509–518.

Krebs, C.J. (1989). *Ecological Methodology*. HarperCollins, New York.

Kristiansen, H.R. (1997). Vulnerability-size effects of electric fishing on population estimate, size distribution and mean weight of a sea trout, *Salmo trutta* L., stock. *Fisheries Management and Ecology* **4**, 179–188.

Kruse, C.G., Hubert, W.A. & Rahel, F.J. (1998). Single-pass electrofishing predicts trout abundance in mountain streams with sparse habitat. *North American Journal of Fisheries Management* **18**, 940–946.

Laffaille, P., Briand, C., Fatin, D., Lafage, D. & Lasne, E. (2005). Point sampling the abundance of European eel (*Anguilla anguilla*) in freshwater areas. *Archive Hydrobiology* **162**, 91–98.

Lamarque, P. (1965). Situation actuelle de la peche electrique en France reserches sun ses bases neurophysioliques. Report to: *Commission Européenne consultative pour les pêches interierures.*

Lamarque, P. (1967). Electrophysiology of fish subject to the action of an electric field. In: Vibert, R. (ed.) *Fishing with Electricity: Its Application to Biology and Management*: 65–89. Fishing News Books Limited, Farnham, Surrey.

Lamarque, P. (1990). Electrophysiology of fish in electric fields. In: Cowx, I.G. & Lamarque, P. (eds.) *Fishing with Electricity: Applications in Freshwater Fisheries Management* Fishing News Books, Blackwell Scientific Publications. Pages 4–33.

Lambert, P., Feunteun, E. & Rigaud, C. (1994). Etude de l'anguille en marais d'eau douce. Première analyse des probabilités de capture observées lors des inventaires par pêche électrique. *Bulletin Français de la Pêche et de la Pisciculture* **334**, 111–121.

Larson, E.I., Meyer, K.A. & High, B. (2014). Incidence of spinal injuries in migratory Yellowstone cutthroat trout captured at electric and waterfall velocity weirs. *Fisheries Management and Ecology* doi:10.1111/fme.12100

Lazauski, H.G. & Malvestuto, S.P. (1990). Electric fishing: results of a survey on use, boat construction, configuration and safety in the USA. In: Cowx, I.G. (ed.) *Developments in Electric Fishing*: Fishing News Books, Blackwell Scientific Publications. Pages 327–340.

Leslie, P.H. & Davis, D.H.S. (1939). An attempt to determine the absolute number of rats on a given area. *Journal of Animal Ecology* **8**, 94–113.

Lethlean, N.G. (1953). An investigation into the design and performance of electric fish screens and an electric fish counter. *Transactions of the Royal Society, Edinburgh* **62**, 479–562.

Liu, Qi-Wen (1990). Development of the model SC-3 alternating current scan fish driving device. In I. Cowx (ed.), Developments in electric fishing. *Proceedings of an International Symposium on Fishing with Electricity*, Hull, UK, 1988. Oxford: Fishing News Books, Blackwell Scientific Publications. Pages 46–50.

Loeb, J., & Maxwell, S.S. (1896). Zur Theorei des Galvanotropismus *Pflügers Archiv – European Journal of Physiology* **63**, 121–144.

Lui, Q-Z., Wu, D., Xu, R. & Li, J. (1990). A method for improving fishing efficiency in lakes by using a seine net with pulsed current. In: Cowx, I.G. (ed.) *Developments in Electric Fishing*: 41–46. Wiley-Blackwell.

Mach, E. (1875). Umfallen von Fischen gegen den negative Pol; elektrische Betaubung der Versuchstiere *Grundlinien der Lehre von den Bewegungs-Empfindungen*. Leipzig.

Mackereth, F.J.H., Heron, J. & Talling, J.F. (1978). *Water Analysis: Some Revised Methods for Limnologists*. Scientific Publication Freshwater Biological Association (UK) no. 36. Freshwater Biological Association, Ambleside, Cumbria.

Maletzky, B.M. (1981). *Multiple Monitored Electroconvulsive Therapy*. CRC Press, Boca Raton, FL.

Malvestuto, S.P. and Sonski, B.J. (1990). Catch rate and stock structure: a comparison of daytime versus night-time electric fishing on West Point Reservoir, Georgia, Alabama. . In: Cowx, I.G. (ed.) *Developments in Electric Fishing*:. Fishing News Books, Blackwell Scientific Publications, pages 210–218

Mann, R.H.K. & Penczak, T. (1984). The efficiency of a new electrofishing technique in determining numbers in a large river in central Poland. *Journal of Fish Biology* **24**, 173–185.

Marking, L.L. and Meyer, F.P. (1985). Are better fish anaesthetics needed in fisheries, *Fisheries* **10**, 6, 2–5.

May, O. (1911). The response of normal and abnormal muscle to Leduc's interrupted current. DOI http://dx.doi.org/10.1093/brain/34.2-3.272

McFarland, W.N. (1959). The use of anaesthetics for the handling and the transport of fishes. *California Fish and Game* **46**, 407–431.

McLain, A. (1957). The Control of the Upstream Movement of Fish with pulsated Direct Current, *Transactions of the American Fisheries Society* **86**, 269–284

McMichael, G.A. (1993). Examination of electrofishing injury and short-term mortality in hatchery rainbow trout. *North American Journal of Fisheries Management* **13**, 229–233.

Meador, M.R. (2005). Single-pass versus two pass boat electrofishing for characterising river fish assemblages: Species richness estimates and sampling distance. *Transactions of the American Fisheries Society* **134**, 59–67.

Mesa, M.G. & Schreck, C.B. (1989). Electrofishing mark-recapture and depletion methodologies evoke behavioural and physiological changes in cutthroat trout. *Transactions of the American Fisheries Society* **118**, 644–658.

Meyer, K.A. & High, B. (2011). Accuracy of Removal Electrofishing Estimates of Trout Abundance in Rocky Mountain Streams. *North American Journal of Fisheries Management* **31**, 923–933.

Miranda, L.E. & Dolan, C.R. (2004). Electrofishing power requirements in relation to duty cycle. *North American Journal of Fisheries Management* **24**, 55–62.

Miranda, L.E. & Kratochvíl, M. (2005). Boat electrofishing relative to anode arrangement. *Transactions of the American Fisheries Society* **137**, 1358–1362.

Mishelovich, G.M. & Aslanov, G.A. (1990). Multi-electrode systems of electric fishing. In: Cowx, I.G. (ed.) *Developments in Electric Fishing*: 281–296. Wiley-Blackwell.

Mitton, C.J.A. & McDonald, D.G. (1994). Consequences of pulsed-DC electrofishing and air exposure to rainbow trout (*Oncorhynhus mykiss*). *Canadian Journal of Fisheries and Aquatic Sciences* **51**, 1791–1798.

Monan, G.E. & Engstrom, D.E. (1962). Development of a mathematical relationship between electric-field parameters and the electrical characteristics of fish. *Fishery Bulletin* **63**, 123–136.

Monbiot, G. (2015). We should be outraged by Europe slaughtering sea life in the name of 'science' http://www.theguardian.com/environment/georgemonbiot. 9 Feb 2015

Muth, R.T. & Ruppert, J.B. (1997). Effects of two electrofishing currents on captive ripe razorback suckers and subsequent egg-hatching success. *North American Journal of Fisheries Management* **16**, 473–476.

Nagel, W.A. (1895). Uber Galvanotaxis. *Pflügers Archive f.d. ges. Physiology* **59**, 603–642.

Nakamura, K. (1992). Effect of precooling on cold air tolerance of the carp *Cyprinus carpio*. *Bulletin of the Japanese Society of Scientific Fisheries* **58**, 1615–1620.

Nelva, A., Persat, H. & Chessel, D. (1979). Une nouvelle methode d'etude des peuplements ichtyologiques dans les grands cours d'eau par echantillonage ponctuel d'abondance. *C. R. Acad. Sci. Paris* **289**(D), 1295–1298.

Nordgreen, A.H., Hoel, A., Slinde, E., Møller, D. & Roth, B. (2008). Effect of various electric field strengths on current durations on stunning and spinal injuries of Atlantic herring. *Journal of Aquatic Animal Health* **20**, 110–115.

Nordwall, F. (1999). Movements of brown trout in a small stream: effects of electrofishing and consequences for population estimates. *North American Journal of Fisheries Management* **19**, 462–469.

Novotny, D.W. & Priegel, G.R. (1974). Electrofishing boats: improved designs and operational guidelines to increase the effectiveness of boom shockers. *Wisconsin Department of Natural Resources Technical Bulletin*, No. 73, 1–48.

Nye, V. (2014). How electric eels use shocks to 'remote control' other fish. *The Conversation* 4th December 2014.

O'Farrell, M., Burger, C., Crump, R. & Smith, K. (2014). Blocking or guiding upstream-migrating fish: a commentary on the success of the graduated field electric fish barrier. *International Fish Screening Techniques: Proceedings of the International Fish Screening Techniques Conference* 2011, Eds: Turnpenny, A.W.H and Horsfield, R.A. pp165–175 WIT press, Southampton, UK.

Pajos, T.A. & Weise, J.G. (1994). Estimating populations of larval sea lamprey with electrofishing sampling methods. *North American Journal of Fisheries Management* **14**, 580–587.

Paragamian, V.L. (1989). A comparison of day and night electrofishing: size structure and catch per unit effort for smallmouth bass. *North American Journal of Fisheries Management* **9**, 500–503.

Penczak, T. (1985). Influence of site area on the estimation of the density of fish populations in a small river. *Aquaculture and Fisheries Management* **1**, 273–285

Penczak, T., Agostinho, A.A., Glowacki, L. & Gomes, L. (1997). The effect of artificial increases in water conductivity on the efficiency of electric fishing in tropical streams (Paraná, Brazil). *Hydrobiologia* **350**, 189–202.

Perrow, M.R., Jowitt, A.J.D. & Zambrano González, L. (1996). Sampling fish communities in shallow lowland lakes: point-sample electric fishing *vs.* electric fishing within stop-nets. *Fisheries Management and Ecology* **3**, 303–313.

Persat H. & Copp, G.H. (1990). Electric fishing and point abundance sampling for the ichthyology of large rivers. In Cowx, I.G. (ed.), *Development in electric fishing*. Oxford: Blackwell Scientific Publications Ltd; 1990, 197–209.

Peterson, J.T., Thurow, R.F., & Guzevich, J.W. (2004). An evaluation of multipass electrofishing for estimating the abundance of stream-dwelling salmonids. *Transactions of the American Fisheries Society* **133**, 462–475.

Peterson, J.T., Banish, N.P. & Thurow, R.F. (2005). Are block nets necessary? Movement of stream-dwelling salmonids in response to three common survey methods. *North American Journal of Fisheries Management* **25**, 732–743.

Peterson, N.T. & Cederholm, C.J. (1984). A comparison of the removal and mark-recapture methods of population estimation for juvenile coho salmon in a small stream. *North American Journal of Fisheries Management* **4**, 99–102

Pickering, A.D., Pottinger, T.G. & Christie, P. (1982). Recovery of the brown trout, *Salmo trutta* L., from acute handling stress: a time-course study. *Journal of Fish Biology* **20**, 229–244.

Pickering, A.D. (1993). Growth and stress in fish production. *Aquaculture* **111**, 51–63.

Pierce, R.B., Coble, D.W. & Corley, S.C. (1985). Influence of river stage on shoreline electrofishing catches in the upper Mississippi River. *Transactions of the American Fisheries Society* **114**, 857–860.

Pugh, L.L. & Schramm, H.L. Jr. (1998). Comparison of electrofishing and hoopnetting in lotic habitats of the lower Mississippi River. *North American Journal of Fisheries Management* **18**, 649–656.

Pusey, B.J., Kennard, M.J., Arthur, J.M. & Arthington, A.H. (1998). Quantitative sampling of stream fish assemblages: single- vs multiple-pass electrofishing. *Journal of Australian Ecology* **23**, 365–374.

Readman, G.D., Owen, S.F., Murrell, J.C. & Knowles, T.G. (2013). Do fish perceive anaesthetics as aversive? *PLoS ONE* **8**(9), e73773. doi:10.1371/journal.pone.0073773

Regis, J., Pattee, E. & Lebreton, J.D. (1981). A new method for evaluating the efficiency of electric fishing. *Archiv für Hydrobiologie* **93**, 68–82.

Reynolds, J.B. & Holliman, F.M. (2002). *Guidelines for Assessment and Reduction of Electrofishing-induced Injuries in Trout and Salmon*. Shocking News. A Review Paper Presented at Wild Trout VII. School of Fisheries and Ocean Sciences, University of Alaska Fairbanks, Fairbanks, U.S.A.

Reynolds, J.B., Kolz, A.L., Sharber, N.G. & Carothers, S.W. (1988). Comments: electrofishing injury to large rainbow trout. *North American Journal of Fisheries Management* **8**, 516–518.

Ricker, W.E. (Ed) (1968). Methods for the Assessment of Fish Production in Fresh Waters. *IBP Handbook No.3*. Blackwell Scientific Publications.

Ricker, W.E. (1975). Computation and interpretation of biological statistics of fish populations. *Fisheries Research Board of Canada, Bulletin 191*.

Riley, S.C. & Fausch, K.D. (1992). Underestimation of trout population size by maximum likelihood removal estimates in small streams.

Roach, S.M. (1999). Influence of electrofishing on the mortality of Arctic Grayling eggs. *North American Journal of Fisheries Management* **19**, 923–929.

Robson, D.S. & Regier, H.A. (1964). Sample size in Petersen mark-recapture experiments. *Transactions of the American Fisheries Society* **93**, 3, 215–226.

Robson, D.S. & Regier, H.A. (1968). Estimation of population number and mortality rates. In: *Methods for assessment of fish production in freshwaters* (Ed. W.E. Ricker) I.B.P. Handbook No. 3, pp 124–138. Blackwell Scientific publications, Oxford.

Robson, D.S. & Spangler, G.R. (1978). Estimation of Population Abundance and Survival In: *Ecology of Freshwater Fish Production* (Ed. S.D. Gerking) pp26–52 Blackwell Scientific Productions.

Rose, J.D. (2002). The Neurobehavioral Nature of Fishes and the Question of Awareness and Pain. *Reviews in Fisheries Science* **10**, 1–38.

Rosenberger, A.E. & Dunham, J.B. (2005). Validation of abundance estimates from mark-recapture and removal techniques for rainbow trout captured by electrofishing in small streams. *North American Journal of Fisheries Management* **25**, 1395–1410.

Ross, L. & Ross, B. (1999). *Anaesthetic and Sedative Techniques for Aquatic Animals*, 2nd ed. Blackwell Science, Oxford.

Ruppert, J.B. & Muth, R.T. (1997). Effects of electrofishing fields on captive juveniles of two endangered cyprinids. *North American Journal of Fisheries Management* **17**, 314–320.

Sammons, S.M. & Bettoli, P.W. (1999). Spatial and temporal variation in electrofishing catch rates of three species of black bass (*Micropterus* spp.) from Normandy Reservoir, Tennessee. *North American Journal of Fisheries Management* **19**, 454–461.

Sanders, R.E. (1992). Day versus night electrofishing catches from near-shore waters of the Ohio and Muskingum rivers. *Ohio Journal of Science* **92**, 51–59.

Sattari, A., Mirzargar, S.S., Abrishamifar, A., Lourakzadegan, R., Bahonar, A., Mousavi, H.E. & Niasari, A. (2009). Comparison of electroanesthesia with chemical anesthesia (MS222 and clove oil) in rainbow trout (*Oncorhynchus mykiss*) using plasma cortisol and glucose responses as physiological stress indicators. *Asian Journal of Animal and Veterinary Advances* **4**, 306–313.

Schill, D.J. & Beland, K.F. (1995). Electrofishing injury studies: a call for population perspective. *Fisheries* **20**, 38–29.

Schill, D.J. & Elle, S.F. (2000). Healing of electroshock-induced hemorrhages in hatchery rainbow trout. *North American Journal of Fisheries Management* **20**, 730–736.

Scholten, M. (2003). Efficiency of point abundance sampling by electro-fishing modified for short fishes. *Journal of Applied Ichthyology* **19**, 265–277.

Scruton, D.A. & Gibson, R.J. (1995). Quantitative electrofishing in Newfoundland and Labrador: Result of workshops to review current methods and recommend standardization of techniques; *Canadian Manuscript Report Fisheries & Aquatic Science St. John's, NF (Canada)*, 20–22 Apr 1993 152 pp.

Schwartz, J.S. & Herricks, E.E. (2004). Use of Prepositioned Areal Electrofishing Devices with Rod Electrodes in Small Streams. *North American Journal of Fisheries Management* **24**, 1330–1340.

Seager, J., Milne, I., Mallett, M. & Sims, I. (2000). Effects of short-term oxygen depletion on fish. *Environmental Toxicology & Chemistry* **19**, 12 pp 6.

Seber, G.A.F. & Le Cren, E.D. (1967). Estimating population parameters from catches large relative to the population. *Journal of Animal Ecology* **36**, 631–643.

Sharber, N.G. & Carothers, S.W. (1988). Influence of electrofishing pulse shape on spinal injuries in adult rainbow trout. *North American Journal of Fisheries Management* **8**, 117–122.

Sharber, N.G., Carothers, S.W., Sharber, J.P., De Vos, J.C., Jr. & House, D.A. (1994). Reducing electrofishing-induced injury of rainbow trout. *North American Journal of Fisheries Management* **14**, 340–346.

Sharber, N.G., Carothers, S.W., Sharber, J.P., De Vos, J.C.Jr. & House, D.A. (1995). Reducing electrofishing-induced injury of rainbow trout: response to comment. *North American Journal of Fisheries Management* **15**, 965–968.

Sharber, N.G. & Black, J.S. (1999). Epilepsy as a unifying principle in electrofishing theory: a proposal. *Transactions of the American Fisheries Society* **128**, 666–671.

Simonson, T.D. & Lyons, J. (1995). Comparison of catch per unit effort and removal procedures for sampling stream fish assemblages. *North American Journal of Fisheries Management* **15**, 2, 419–427.

Smith, C.E. (1978). Transportation of salmonid fishes. In: *Manual of Fish Culture*. U.S. Fish & Wildlife Service, Washington, DC.

Smolian, K. (1944). Die Electrofischerei. *Sammlung fischerilicher Zeitfragen Herausgegeben von Reichsverband der Deutschen Fischerie*, Heft 35. Verlag J. Neumann, Neudamm und Berlin.

Snyder, D.E. (1995). Impacts of electrofishing on fish. *Fisheries* **20**, 26–27.

Snyder, D.E. (2003). Electrofishing and its harmful effects on fish, *Information and Technology Report* USGS/BRD/ITR-2003-0002: US Government Printing Office, Denver, CO, 149pp

Soetaert, M., Decostere, A., Polet, B. & Chiers, K. (2015). Electrotrawling: a promising alternative fishing technique warranting further exploration *Fish and Fisheries* **16**, 104–124.

Solomon, D.J. (1999). Addressing the problems of damage to fish caused by electric fishing. In: *Report to UK Environment Agency*. UK Environment Agency, London.

Solomon, D.J. & Hawkins, A.D. (1981). Fish capture and transport. *Aquarium Systems* **1981**, 197–221.

Sternin, V.G. Nikonorov, I.V. & Bumeister, Y.K. (1976). Electrical fishing, theory and practice [English translation of Sternin *et al.* 1972 from Russian by E. Vilim]. In: *Israel Program for Scientific Translations*, Keter Publishing House Jerusalem Ltd, Jerusalem.

Stewart, P.A.M. (1990). Electrified barriers for marine fish. In: Cowx, I.G. (ed.) *Developments in Electric Fishing*. Fishing News Books, Blackwell Scientific Publications. Pages 243–255.

Stott, B. & Russell, I.C. (1979). An estimate of a fish population which proved to be wrong. *Fisheries Management* **10**, 169–170.

Taylor, A.L. & Solomon, D.J. (1979). Critical factors in the transport of live freshwater fish. The use of anaesthetics as tranquillizers. *Fisheries Management*, **10**, 153–157.

Taylor, N.I. & Ross, L.G. (1988). The use of hydrogen-peroxide as a source of oxygen for the transportation of live fish. *Aquaculture* **70**, 1–2.

Thompson, K.G., Bergersen, E.P. & Nehring, R.B. (1997). Injuries to brown trout and rainbow trout induced by capture with pulsed direct current. *North American Journal of Fisheries Management*, **17**, 141–153.

Twedt, D.J., Guest, W.C. & Farquhar, B.W. (1992). Selective dipnetting of largemouth bass during electrofishing. *North American Journal of Fisheries Management*, **12**, 609–611.

Tytler, P. & Hawkins, A.D. (1981). Vivisection, anaesthetics and minor surgery. A.D.Hawkins (ed) In: *Aquarium Systems* Academic Press, London

Vandergoot, C.S., Murchie, K.J., Cooke, S.J., Dettmers, J.M., Bergstedt, R.A. & Fielder, D.G. (2011). Evaluation of two forms of electroanesthesia and carbon dioxide for short-term anesthesia in walleye. *North American Journal of Fisheries Management* **31**, 5, 914–922

VanderKooi, S.P., Maule, A.G. & Schreck, C.B. (2001). The effects of electroshock on immune function in juvenile spring Chinook salmon. *Transactions of the American Fisheries Society*, **130**, 397–408.

Van Zee, B.E., Hill, T.D. & Willis, D.W. (1996). Comment: clarification of the outputs from a Coffelt VVP-15 Electrofisher. *North American Journal of Fisheries Management* **16**, 477–478.

Vehanen, T., Sutela, T., Jounela, P., Huusko, A. & Mäki-Petäys, A. (2012). Assessing electric fishing sampling effort to estimate stream fish assemblage attributes. *Fisheries Management and Ecology*. doi:10.1111/j.1365-2400.2012.00859.x

Verheijen, F.J. & Flight, W.G.F. (1992). What we may and may not do to fish. *Report to Eurogroup for Animal Welfare, Brussels.* UBI-M-92.CP-074. 6 pp.

Vibert, R. (1963). Neurophysiology of electric fishing. *Transactions of the American Fisheries Society* **92**, 265–275

Vibert, R. (ed.) (1967). *Fishing with Electricity: Its Application to Biology and Management*. Fishing News Books Limited, Farnham, Surrey.

Vibert, R. (1967). Applications of electricity to inland fishery biology and management. In: Vibert, R. (ed.) *Fishing with Electricity: Its Application to Biology and Management*: 3–50. Fishing News Books Limited, Farnham, Surrey.

Vincent, R. (1971). River electrofishing and fish population estimates. *The Progressive Fish-Culturist* **33**, 163–169.

Waring, C.P., Stagg, R.M. & Poxton, M.G. (1992). The effect of handling on flounder (*Platichthys flesus* L.) and Atlantic salmon (*Salmo salar* L.). *Journal of Fish Biology* **41**, 131–144.

Warry, F.Y., Reich, P., Hindell, J.S., McKenzie, J. & Pickworth, A. (2013). Using new electrofishing technology to amp-up fish sampling in estuarine habitats. *Journal of Fish Biology* **82**, 1119–1137

Weddle, G.K. & Kessler, R.K. (1993). A square-metre electrofishing sampler for benthic riffle fishes. *Journal of North American Benthological Society* **12**, 291–301.

Weisser, J.W. & Klar, G.T. (1990). Electric fishing for sea lampreys (*Petromyzon marinus*) in the Great Lakes region of North America. In: I.G. Cowx (ed.), *In Developments in Electric Fishing*, 59–64. Fishing News Books, Blackwell Scientific Publications.

Welton, J.S., Beaumont, W.R.C. & Mann, R.H.K. (1990). The use of boom-mounted multi-anode electric fishing equipment for a survey of the fish stocks of the Hampshire Avon. In: I.G. Cowx (ed.) *Developments in Electric Fishing*, 236–242. Fishing News Books, Blackwell Scientific Publications, Oxford.

Whaley, R.A., Maughan, O.E. & Wiley, P.E. (1978). Lethality of electroshock to two freshwater fishes. *The Progressive Fish-Culturist* **40**, 161–163.

Whitney, L.V. & Pierce, R.L. (1957). Factors controlling the input of electrical energy into a fish (*Cyprinus carpio* L.) in an electric field. *Limnology & Oceanography* **11**, 55–61.

Wiley, M.L. & Tsai, C.F. (1983). The relative efficiencies of electrofishing vs seines in Piedmont streams of Maryland. *North American Journal of Fisheries Management* **3**, 243–253.

Williams, G.I. (1984). A medium-power electric fishing apparatus. *Fisheries Management* **15**, 169–176.

Winstone, A.J., & Solomon, D.J. (1976). The use of oxygen in the transport of fish *Fisheries Management* **7**, 30–33.

Wood, C.M., Turner, J.D. & Graham, M.S. (1983). Why do fish die after severe exercise? *Journal of Fish Biology* **22**, 189–201.

Wyatt, R.J. (2002). Estimating riverine fish population size from single- and multiple-pass removal sampling using a hierarchical model. *Canadian Journal of Aquatic Science* **59**, 695–706

Zalewski, M. (1983). The influence of fish community structure on the efficiency of electrofishing. *Fisheries Management*, **14**, 177–186.

Zippin, C. (1958). The removal method of population estimation. *Journal of Wildlife Management* **22**, 82–90.

Zhong, W.G. (1990). Model LD-1 electric fishing screen used in reservoir fisheries in China. In: Cowx, I.G. (ed.), *Developments in Electric Fishing*. Fishing News Books, Blackwell Scientific Publications. Pages 297–305.